fifty Fathoms

Statistics Demonstrations for Deeper Understanding

Tim Erickson

Author: Tim Erickson

Design and Layout: eeps media

Good Advice: Michael Allwood, Caroline Ayers, Jill Binker, Kristen Clegg, Bill Finzer, Meg Holmberg, Paul L. Myers, Chris Olsen, Tony Thrall, Joshua Zucker

eeps media is the educational publishing imprint of Epistemological Engineering ("helping you know how you know what you know...since 1987"). We are consultants, writers, and curriculum developers in mathematics, science, technology, and conflict resolution.

For additional information and volume sales, please contact us:

eeps media
5269 Miles Avenue
Oakland, CA 94618–1044
510.653.3377
510.428.1120 fax
publications@eeps.com
http://www.eeps.com

For single copies, call us (510.653.3377) or Key Curriculum Press (800.995.MATH)

Printed in the United States of America, at VIP Litho, San Francisco, CA. http://www.viplitho.com.

ISBN 0-9648496-2-3

8 7 6 5 4 3 10 09 08 07 06 05

First Printing, September 2002
Third Printing, July 2005

The cover: Photo © Randy Morse, used with permission.

Gary, the fish, is a *Hypsypops rubicundus* living in Southern California. We are aware that Randy took his picture at more like ten fathoms, not fifty, but there is not enough light down there. You can see more of Randy's work at http://www.goldenstateimages.com.

System requirements: Macintosh® PPC running OS8, Pentium® running Windows® 95 or later. You need 8MB of memory. You must have a CD-ROM drive.

Contents

What Is This Book About?

This is a book of computer-based demonstrations of concepts that appear in many introductory statistics courses. They use the power of Fathom Dynamic Statistics™ software to clarify complicated statistical ideas. For example, if you're studying least-squares linear regression, it really helps to wiggle the line around on a diagram like this one:

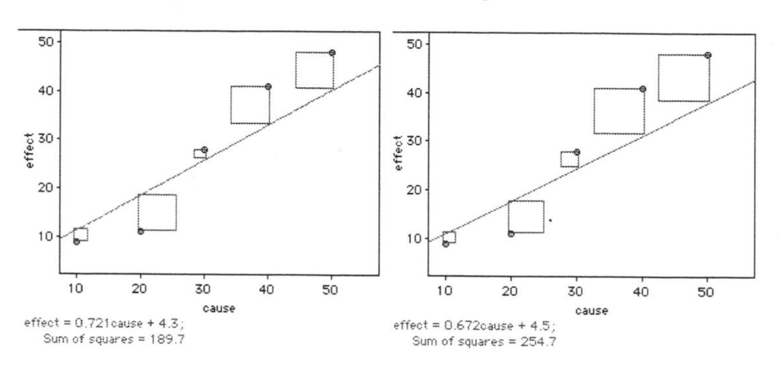

and try to minimize the "sum of squares" number, which is the total area of all the squares on the diagram. Before this kind of technology, you might have just believed the regression line some computer or calculator gave you. If you wanted to get to the root of matters, you had to grapple with a formula such as:

$$Y = \bar{y} + r_{xy}\left(\frac{\sigma_y}{\sigma_x}\right)(X - \bar{x}),$$

which is a great formula, but not particularly accessible.

Whether you understand the formula or not, however, seeing what it *looks like* to minimize the sum of squares will make you a more sophisticated user of least-squares lines. Then, when the computer gives you a least-squares line, you will better understand what it is, what it is not, and whether it's what you really want.

The topics in the book range from first-week material such as the median and the mean to quite sophisticated concepts such as the power of a statistical test and the dire consequences of heteroscedasticity.

How to Use This Book

Fifty Fathoms is for teachers and learners of statistics.[1]

If you're a teacher, the most straightforward way to use these demos is as whole-class presentations. You perform these demos for the class—using some projection system—either as an introduction to a topic or as review, perhaps clarifying some idea students have trouble with, or simply showing them the concepts in a different way.

You need no experience with Fathom to do these demos—the hard stuff is mostly done for you in the files on the CD. The book gives you step-by-step instructions pitched to the courageous novice—someone who is experienced with computers but not with this software.

1. And, as all stats teachers know, we are learners as well: $T \cap L = T$.

On the other hand, if you want additional background and some great tips, read *and do* "Tutorial: An Extended Example" on page 13, and look at "A Few Good Skills" on page 17 and "Fathom Overview" on page 19.

Ideally, you should make these demos as interactive as possible: ask students to predict what will happen if you move this point or that slider; have them sketch what they think the next graph will look like; have them discuss with their neighbor the meaning of some value. There are questions within the demos to give you ideas about what to ask, but you will find many opportunities throughout each demo for some give-and-take with the class.

That said, the first time, you may want to stay quite close to the book and not open things up too far. But as you get more comfortable with the software and the material, you can deviate quite widely from the particular path each demo takes.

If you want to make extra copies of these book pages or the computer files, be sure to get permission; write to **publications@ eeps.com**.

Using These Demos As Learning Guides. You can use these outside of a lecture, however. For example, assign one of these demos for your students to do in groups—at a computer in the back of the class or in the library, say—before you introduce a topic. These demos can also be good for independent, "self-starter" students or as inspirations for projects. In general, however, these are too hard for *independent* work by students learning a statistics topic *for the first time*. This is a book of demos: they supplement *but do not substitute for* a group, a teacher, or a good textbook.

Still, these demos can be great as a reinforcement for learners—such as ourselves—who are *relearning* stats topics. For example, if you have just been told that you're teaching statistics next semester and you're feeling a little rusty, these can help. Or these may be perfect for students at the end of the semester, reviewing things they already "know."

If You Are a Student. Use these demos to review material you have already learned, whether you need to remember things you have forgotten or just want to be more secure in your knowledge.

If you are learning these concepts for the first time, you will probably need a textbook, a group, a teacher, or some combination of these, to make best use of these demos. Some of these demos go beyond the typical first course in statistics, so don't be alarmed if a demo seems to be about something you've never heard of (e.g., the Cauchy distribution).

About the Questions

Most of the demos have explicit questions in the text. Some of these questions are easy and just serve to clarify what's going on. Some are very hard; some are practically projects in themselves. They appear more or less in order of difficulty. These questions serve at least three different purposes:

1. If you're teaching a class and using these demos with a projector, the easier questions will help you check for understanding; harder ones can be starting points for discussion.
2. If you're teaching a class and assigning demos for students to work through, pick some questions for students to write up and hand in.
3. If you're learning on your own, first try simply to *understand* the questions, then see which ones you can answer.

Solutions. Many of the questions are investigatory and open-ended; you can approach them in many ways. Rather than promote "book-bloat" with pages of solutions (including "answers will vary") that might not fit the path you want to take, we've decided to put responses to selected questions on the web. If you see this thing: web in the side margin, that means that there's a solution on the web at the time of publication. This way, we can update and expand our solutions, discuss topics in more detail, even add new demos—with your input. If you come up with better solutions to a question, or solutions to questions we haven't yet answered, send them to **answers@eeps.com**. Visit **http://www.eeps.com** to see them.

About Context

The best statistics education uses real data from engaging contexts to illustrate statistical principles. But sometimes, when you need to be general or abstract, reality gets in the way. And a book like this cannot predict what context will be best for *you*. Thus these demos are abstract. We encourage you to endow them with context as you see fit.

For example, in "The Meaning of Mean" on page 22—the very first demo—we use two small sets of pretty-much-arbitrary data. Once you see what the demo is about, however, you could substitute data from the class recorded for some other purpose.

Software, Installation, and Your License

With this book, you get a CD-ROM. The CD-ROM works with both Windows and Macintosh. Put it in your CD-ROM drive. You will see:

- the **ReadMe** file (which you should read) in two formats,
- a folder called **Fifty Fathoms Files**, and
- a special limited-use edition of Fathom called **FiftyFathoms**.

Do the following:

The limited-use Fathom that comes with this book *is not the same* as the full edition of Fathom. If you have that full edition, *do not use this one*.

If you already own Fathom. Copy the **Fifty Fathoms Files** folder and all its contents to your hard disk. It will probably be convenient to put it inside your **Fathom** folder—the folder your Fathom application is located in—perhaps inside the **Sample Documents** folder. Then remove the CD-ROM and put it in a safe place. You're done! Enjoy! Open the files as you would any other Fathom document.

If you do not already own Fathom. You will be using a limited-use introductory edition of Fathom included on the CD-ROM by special arrangement with Key Curriculum Press. *This version requires that the CD-ROM be in your computer when you use it.*

Double-click the **FiftyFathoms** application (the icon that looks like a gold ball) to start the program. Choose **Open** from the **File** menu (on the Mac, this is done for you automatically). In the dialog box that appears, navigate to the **Fifty Fathoms Files** folder on the CD, and choose the file you want to open. After you have done this once, you can also open the files simply by double-clicking them.

Questions about Installation and Use

Can I copy the files to my hard disk?

Yes. But it may be best not to copy the application **FiftyFathoms** itself. If you someday buy a full-featured version of Fathom and install it, there's always the chance that your operating system will become confused about which copy of Fathom to run.

How is the limited-use edition different from the full version?

With this limited-use edition, you can use all of these demos exactly as they are written. But you cannot customize the files on the CD, make new ones, import or export data, print, copy, use the help system, or open any other Fathom files. For that you need the full-featured edition: visit **http://www.keypress.com** to order the full-featured Fathom or a site license. And remember:

To use the limited-use edition of Fathom,
you must have the original CD in the drive.

When I double-click a document name, Windows cannot find the application. Why not?

The first time you use the program, launch the Fathom application by double-clicking the **FiftyFathoms** icon. Then choose **Open** from the **File** menu and navigate to the **Fifty Fathom Files** folder. There, find the file you want and open it. After that, Windows should know that **".ftm"** documents are associated with Fathom.

I can't see all the graphs and tables that I can see in the book. I have to keep scrolling around.

Give Fathom as much screen space as possible. On Windows, maximize the application window and the document window within it. Each file is designed to work at a size of 800 by 600 pixels; at that size, you can see everything in the document without scrolling.

What You May and May Not Do

Buying this book with the CD-ROM entitles you to copy the documents in the **Fifty Fathoms Files** folder to *one* computer to be used as a demo computer, and to *one* other computer so you can prepare your demos, e.g., one in the classroom and one at home.

This way, a teacher or solo learner can do the demos. But suppose you're a teacher and you want *all* the students in your class to do the demos themselves? What then?

One solution is to have each student buy a book. We offer discounts to bookstores, schools, and individuals for volume sales. Contact **sales@eeps.com** and check our web site, **http://www.eeps.com**.

If that's too expensive, the first thing you need is a site license for the Fathom software (**http://www.keypress.com**). With a site license, you can install the software on a server or on individual computers, you get additional curriculum materials, and your students get full use of all of Fathom's features. Besides being less expensive than a class set of books, having the software installed is more convenient than keeping track of a class set of the CDs.

If you do opt for a site license, however, we ask that you write for permission to duplicate both the **Fifty Fathoms Files** from the CD and any pages from the book that you want to distribute to your class. Please contact us at **publications@eeps.com** to tell us how much you will be duplicating; we'll make this work for you at a reasonable cost. For information about permission, you can also check our web site, **http://www.eeps.com**.

What about Workshops?

You may duplicate book pages for workshop use; the handout must cite the book and give purchasing information, for example, our web address.

If This Is Your First Time Using Fathom

To use this book, you don't need to be a Fathom expert, because the instructions for each demo tell you exactly what to do. So you could just pick a demo (e.g., "The Meaning of Mean" on page 22) and jump right in. But some people like to get a little background before they begin. This section offers three ways to prepare for using Fathom.

- "Tutorial: An Extended Example" (below) teaches you basic Fathom skills as it guides you through a specific task.
- "A Few Good Skills" on page 17 gives you additional "tricks of the trade" and things to watch out for.
- "Fathom Overview" on page 19 shows you the big picture: how the program is organized.

Tutorial: An Extended Example

In this example, we'll look at random numbers, uniformly distributed in the interval [0, 1]. Then we'll take their mean and create a sampling distribution of that mean. As in the actual book, we will begin with some things already made.

✣ Open the file **Example.ftm**. It will look like the illustration. Every file we ask you to open is in the **Fifty Fathoms Files** folder or in a subfolder called **other files**. This one is in the subfolder:

The first time you use Fathom, double-click the **FiftyFathoms** icon to start the program and choose the **Example** file in the **File>Open** dialog (this appears automatically on the Mac). Later, you can just double-click the file itself.

On Windows, you may need to maximize your window size to get your screen to match the illustrations.

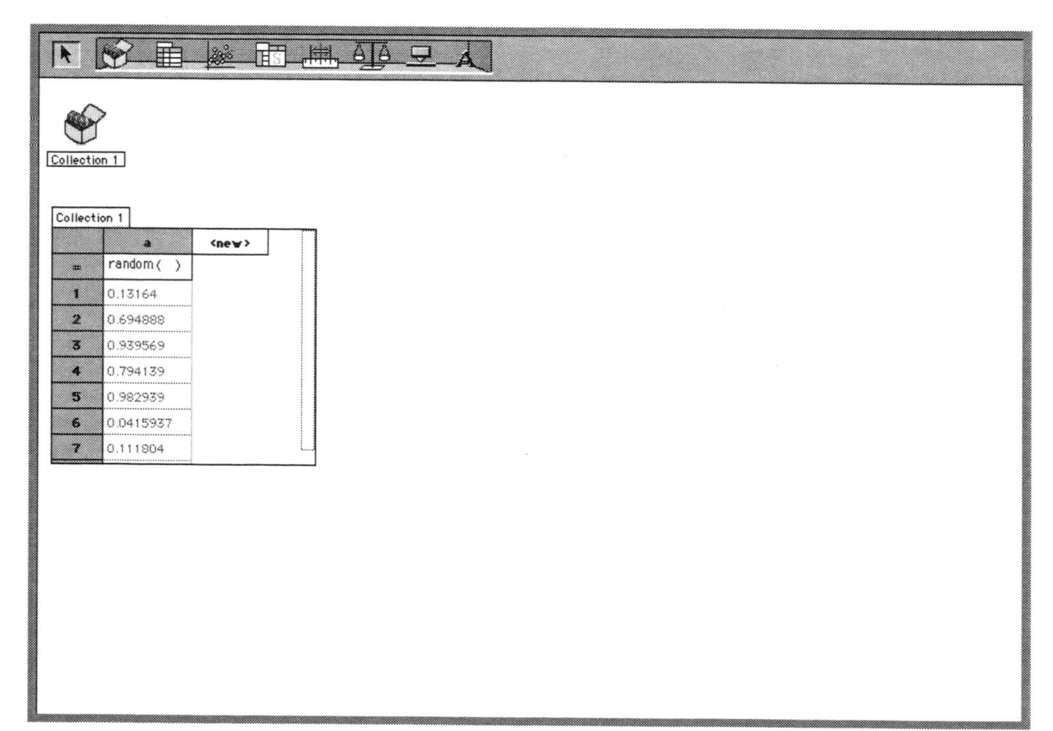

In the upper left, you can see the *collection*, which looks like a box of gold balls. The data in the collection appear in a *case table* just below it. You can see the first seven *cases* in the collection (there are 100 altogether). Each case has one *attribute*, called **a**. The *values* of **a** are determined by a formula, which you can see in the row labeled with an = sign: it's **random( )**.[1]

Now we want to make a *graph* of these data. It's a two-step process.

- Drag a new graph off the *shelf* (see step 1 in the illustration below) and put it in a blank part of your document.
- Drag the attribute **a** to the horizontal axis of the graph. The axis will highlight when you're in the right place. See step 2 in the illustration:

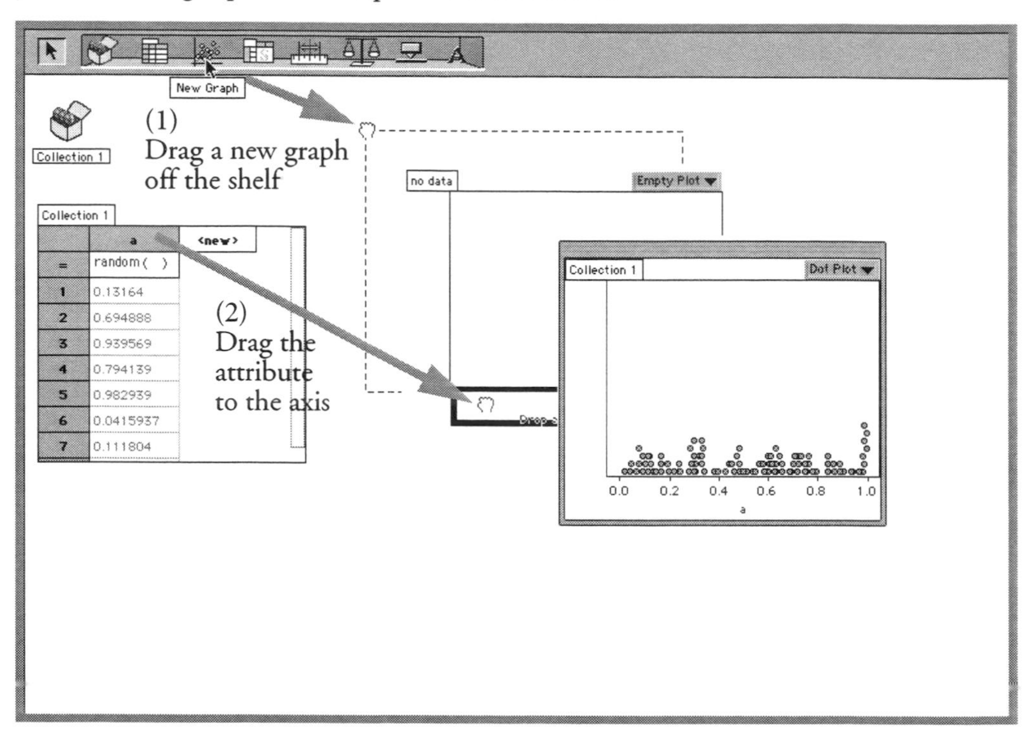

You wind up with a dot plot; you can change the kind of plot with the popup menu in the upper-right corner of the graph.

These 100 values are random, so let's rerandomize them.

- Choose **Rerandomize** from the **Analyze** menu. The shortcut is **clover-Y** on the Mac or **control-Y** in Windows. (In the book, we'll remind you of this shortcut with some text in the margin.) You'll see the points change—both on the graph and in the case table.

Next, let's compute the mean and plot it in the graph.

- Click once on the graph to make sure it is selected. A selected object has a border around it.

1. You will not always see the formulas in the case tables. You can control whether formulas appear by first selecting the case table, and then choosing **Show** (or **Hide**) **Formulas** in the **Display** menu.

Fathom has context menus (see page 19), but for simplicity in the text, we won't remind you of them every time. Use the right button in Windows and **control** with the mouse button on the Mac.

- Choose **Plot Value** from the **Graph** menu—the **Graph** menu in the *menu bar*, not the popup in the graph itself. (Note: if there is no **Graph** menu, that means the graph wasn't selected.) Fathom's *formula editor* appears, labeled **Expression for Value**.

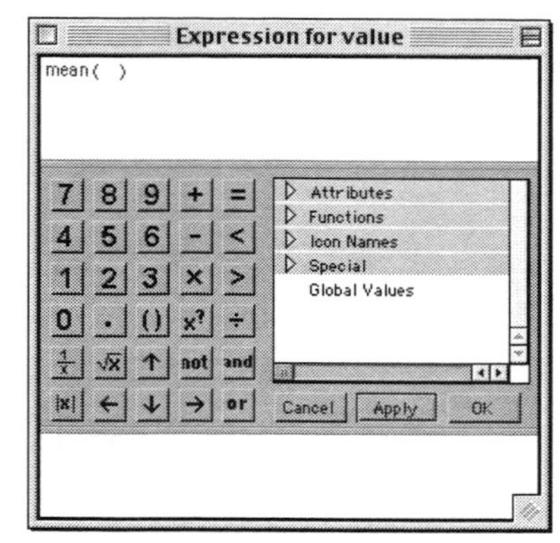

- Enter the formula **mean( )**, as shown in the illustration at right. Notice that when you type the left parenthesis, Fathom adds the right one automatically. Leave the parentheses empty. (We could have used **mean(a)**, but Fathom knows that if you leave it blank, you mean "whatever is on the axis.")
- Press **OK** to leave the formula editor. You'll see the value for the mean at the bottom of the graph and a line showing where that mean is.
- **Rerandomize** again (it's in the **Analyze** menu) to see the mean change.

Creating Measures

Now we want Fathom to collect those means automatically so we can look at the distribution of the values. We can't do this directly from the graph; we need an attribute—but not one that applies to a single case. We will create an attribute that applies to the whole collection. This is a *statistic*; in Fathom, we call it a *measure*.

- Double-click the *collection* (the box with gold balls) to open that collection's *inspector*.

The inspector opens up in a small window. In it, you can control all kinds of things about the collection. There are tabs across the top: **Cases**, **Measures**, **Comments**, and **Display**.

In the files on the CD, we have often done these steps for you; your collection will usually have its measures already defined.

- Click the **Measures** tab to go to that panel. Here is where you define measures. It has three columns: the name of the measure, its value, and its formula.
- Click in the **Measure** column where it says **<new>**.
- Enter **meanA**. This will be the name of our measure.
- Press **return**, **tab**, or **enter** to tell Fathom you're done entering the name.
- Double-click in the **Formula** column opposite your attribute name, as shown. The formula editor opens, labeled **meanA formula**.
- This time, definitely enter **mean(a)**. (There's no axis, so Fathom can't figure out what you want to take the mean of.)
- Press **OK** to leave the editor. The value appears in the inspector; it should be the same as the one you can see in the graph.

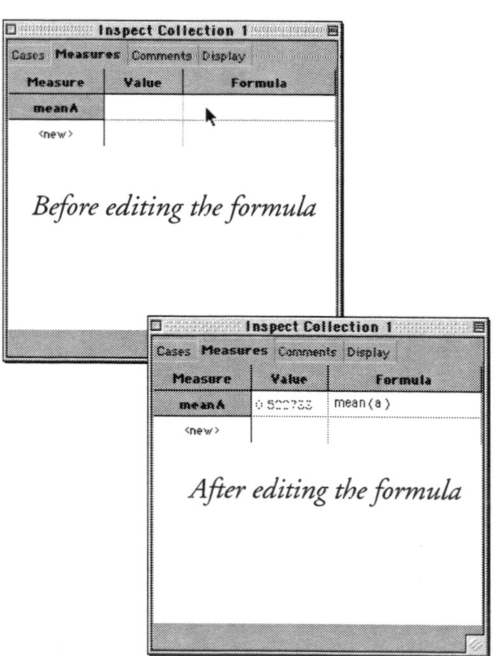

Before editing the formula

After editing the formula

- At this point, though you could just leave that inspector open, it's best to close it (we'll tell you in the book instructions) in order to save screen space—and to avoid confusion, for we're about to create a new collection with its own inspector.

Collecting Measures

Now that you have defined this measure, you can collect many copies of it to study its distribution.

- Click on the collection to select it.
- Choose **Collect Measures** from the **Analyze** menu. Fathom makes a new collection, called **Measures from Collection 1**. Gold balls fly from your first collection to the new one. Let's see what happened.
- Drag a graph from the shelf and put it in some open area.
- Double-click the **Measures from Collection 1** collection to open its inspector (each collection has its own inspector). It should be open to a new panel, called **Collect Measures**. You can see that here is where you control, for example, how many measures Fathom collects—the default is five.
- Click the **Cases** tab to bring up that panel. Now you can see the first case, its attribute, **meanA**, and that value.
- Drag the name of the attribute, **meanA**, to the horizontal axis of the graph. Your window now looks something like this:

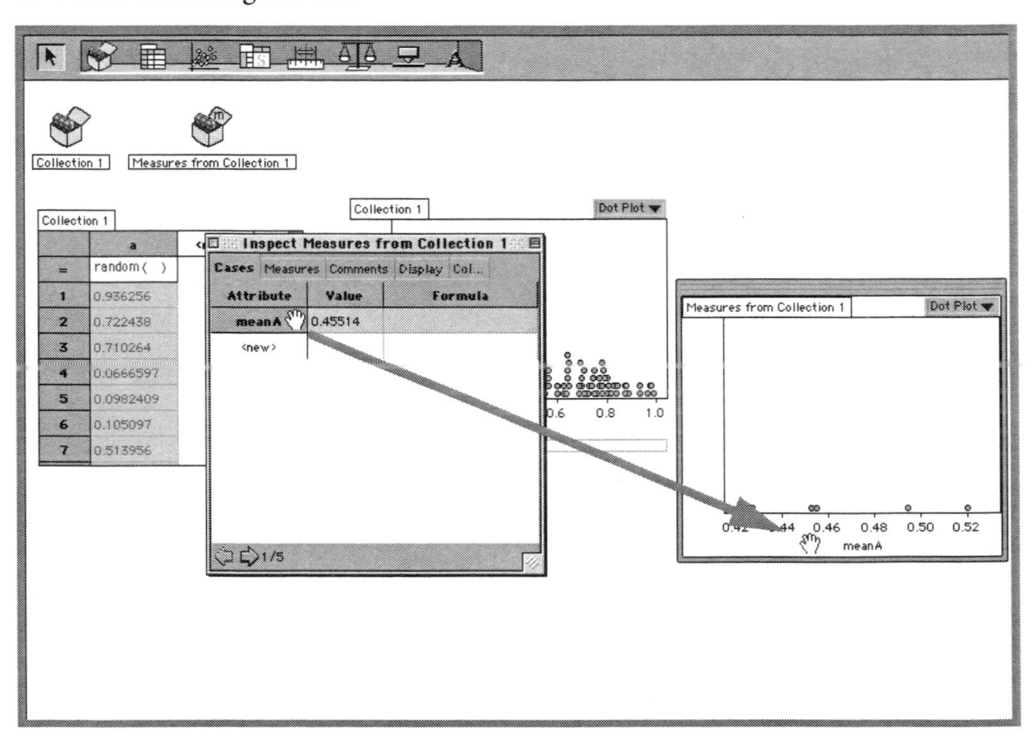

How do you make a case table for the new measures? Just select the box of balls and then drag a new case table off the shelf.

There is an important principle at work that might slip by unnoticed: before, we dragged the attribute name from the case table. This time, we dragged it from the inspector. Either one works; in fact, you can even drag names from other graphs to make a graph.

Also, notice how **meanA**, which was a *measure* in the original collection, is a regular *case attribute* in the measures collection—one you can graph. (If you drag the name of a measure to a graph, nothing happens.) Now let's change how Fathom collects these measures.

- Click the **Collect Measures** tab (it probably says **Col...**) in the inspector to bring up that panel. Notice the **Collect More Measures** button at the bottom.
- Press that **Collect More Measures** button. Fathom collects a new set of five measures.

- Click the **Empty this collection first** box to un-check it. Then press **Collect More Measures**. Note that Fathom collects five *additional* measures, so now your graph has ten points.
- Tired of the flying balls? In the inspector, click the **Animation on** box to *deselect* it and press **Collect More Measures**. Much faster! Also, the graph of the original data—the 100 random numbers—does not update each time.
- Let's get a *lot* of measures. Change the 5 in the box to 200. Press **Collect More Measures**. You'll probably see a progress box showing you how long you'll have to wait. You can stop it wherever it is with the **Cancel** button or by pressing **escape**.

Measures Are Important

The concept of measures is an important—and conceptually deep—idea. In this book, we often use measures (as we did here) to make an approximation to a sampling distribution. Though it has taken a couple of pages to explain how to make a measure and use it, this procedure gets easy with a little practice. And, of course, you will not have to remember the details if you use this book's instructions. Everything is all set up in the documents you will use, and we'll give you specific instructions about what to press when. Even so, the "measures" mechanism is behind many of the cool demos and simulations in this book, so it's probably good that you went though this to see how it's done.

A Few Good Skills

While the above mirrors a typical sequence of instructions, once in a while you have to do something a little tricky, or you make a mistake. And while **Undo** will fix most of your missteps, this section will help you learn some extras that will make your life with Fathom easier and your demos more flexible.

Screen Space Management

One of the greatest things about Fathom is that you can have a jillion graphs and tables all going at once. But each one takes up screen space—and we *all* run out. Mostly we took care of this for you in these demos. But you never know what direction you might go off into, so here are some tips:

- If you have a larger monitor or high-resolution projector, increase your window sizes to the maximum.
- Each type of object in Fathom—case tables, graphs, sliders, statistical tests, collections—behaves the same way: click on it *once* to select it. That gives it a border. Then you can drag the top edge to move it, and *any other edge or corner* to resize it.
- Iconify an object by dragging a corner, shrinking it until it turns into an icon.
- Delete objects you no longer need: **clover-D** on the Mac, the **delete** key in Windows. The version of Fathom that comes with this book will let you delete only objects you made yourself. You can **Hide** objects, however, in all versions. The **Hide** command is in the **Display** menu. And don't forget **Undo**!
- Make case tables *wide* and *short*. After you have looked at the data, you may need case tables only to get at the attribute names.
- If you're really short of space, *delete or hide the case tables*. If you need the attribute names, get them from the **Cases** panel in an inspector. Get the inspector by double-clicking a collection (the box of gold balls).

Rescaling Graph Axes

Dragging. What if a graph does not have the right scale? The basic idea is to grab axis *numbers* (not the axes themselves) and drag them where you need them. Need to zoom in? Push the numbers you don't need off the edge of the graph.

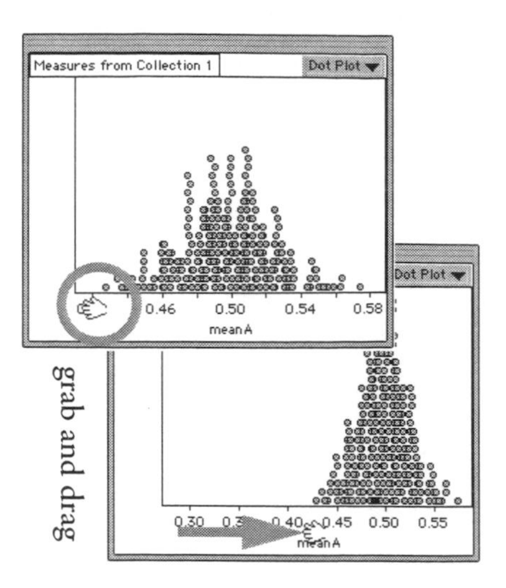

How does it work? When the cursor is over an axis, it turns into a "grabber hand," which changes orientation depending on where you are on an axis. Consider a horizontal axis. If you grab in the middle, dragging just translates the axis. Grab on the right, and it will stretch or shrink from the left edge, and vice versa. It's harder to describe than to do; the illustration shows a graph from the example above where we grab on the left edge and pull to the right.

Zoom tool. If you're pointing at a graph or an axis, and hold down the **option** key on the Mac (**control** in Windows), the cursor changes to a magnifying glass. Click on an axis or graph to zoom in, or drag a rectangle in the graph to zoom to that area. Holding **shift** in addition to **option** or **control** makes you zoom out.

ControlText. For complete control over an axis, double-click the axis numbers. Some *ControlText* appears, in a new object, as shown at right. If you rescale the axes, as described above, the numbers will update. But you can also edit the numbers directly, to get exactly the graph bounds you want. When you're done, press **tab**, **enter**, or **return** to tell Fathom you're done editing.

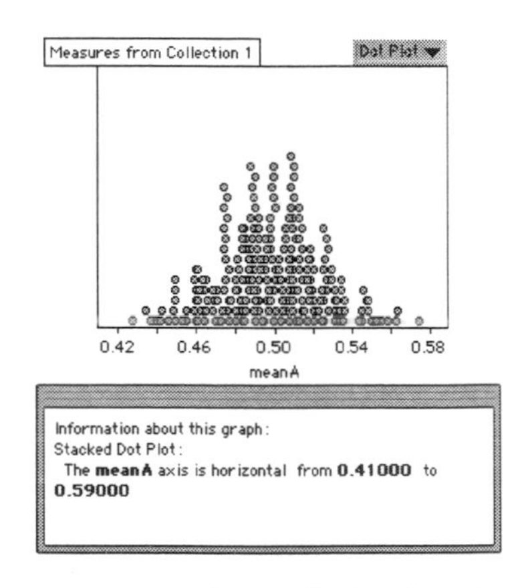

Speed tip: if the cursor is anywhere in that text block, pressing **tab** selects the next editable number. So to rescale this graph to zero-to-one, click in the top line, then press **tab 0 tab 1 tab**—and you're done.

Getting Back to Normal. You have zoomed in, and now you want to show all the data again. Two strategies: re-choose the type of graph (e.g., **Histogram**) from the popup menu in a graph; or select the graph and choose **Rescale Graph Axes** from the **Graph** menu.

Deleting Objects

That ControlText object is great while it's there, but how do you get rid of it—or any other graph, table, or whatever that has outlived its usefulness? Easy. Select it and press **delete** (Windows) or **clover-D** (Mac). If Fathom has trouble, you can also select it and choose **Delete ControlText** (or **Delete** whatever) from the **Edit** menu. Again, with the limited-use edition of Fathom, you can only delete objects you made yourself.

Undo

Undo is your friend. Like its cousin, *The Geometer's Sketchpad*, Fathom supports near-infinite Undo—and Redo—so it's easy to get back to where things had not yet gone wrong. It's in the **Edit** menu; the shortcut is **clover-Z** (Mac) or **control-Z** (Windows). For various existential reasons, actually collecting measures or sampling is not undoable.

Context Menus

If you want to look like a real Fathom Master, use *context menus*. Invoke them with the right button in Windows, or on the Mac by holding down **control** (*not* **clover**!) and using the only mouse button.

Just point at any object, bring up the context menu, and voilà! you get a menu of all sorts of things that are relevant to the object you're pointing at. That means that you don't have to remember which menu the command is in.

For example, you want to add cases to a collection. Point at a collection, or a case table, or a graph, and choose **New Cases...** from the context menu. (**Delete** [object] is on the context menu too.)

The most important, or common, context menu item is usually on top. For example, when you point at an attribute *name* in a case table, the top item is **Edit Formula**.

Fathom Overview

This section is a bird's eye view of Fathom's design. Most people just use the program and absorb this overview by osmosis. But others prefer to read how we designers think the program is organized. If you're one of these, you (like this author) have probably actually *read* the rules of Monopoly. This section is for you:

In Fathom, the data live in *Collections*. Every other place you see data—for example, in case tables or graphs—is a *view* of the data.

Collections look like boxes, usually boxes of little gold balls. If you double-click a collection, its *inspector* appears. An inspector has four or more panels (named **Cases**, **Measures**, **Comments**, etc.) that you get to by clicking on different tabs.

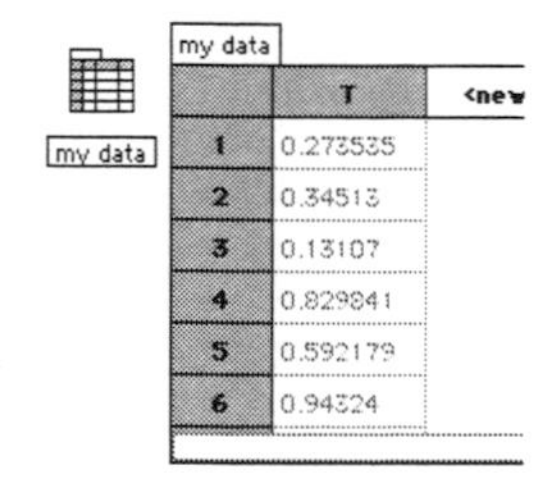

Case tables look like spreadsheets. The column headers—the variable names—are called *attributes*. The rows are called *cases*. So in data about people, each person could be a case; the attributes might be **name**, **height**, **age**, and so forth. Each case—each person—would have a value for each attribute. (In the collection, each case appears by default as a gold ball.)

Graphing is easy. You make an empty graph—drag a new graph off the *shelf*—and then drag the names of attributes (from case tables, inspectors, or other graphs) to the axes of the graph. A popup menu on the graph itself lets you choose the kind of graph. The set of available graphs is determined by the attributes on the axes.

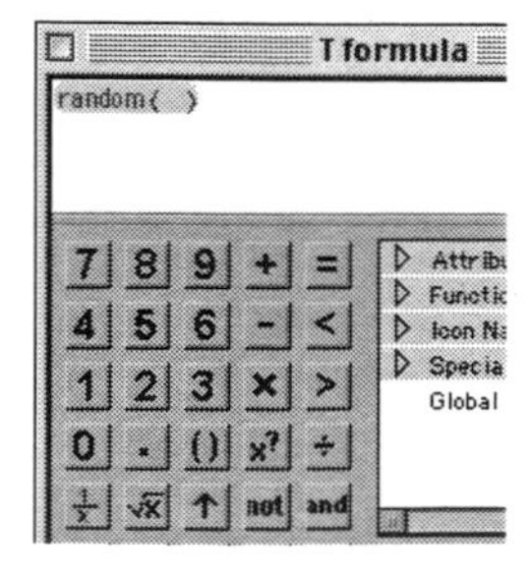

Sometimes, the values for the attributes of a case are regular data that someone entered. But other times—and frequently in this book—Fathom generates the values based on a *formula*. Fathom has a *formula editor* for entering these things. You can use regular calculation and any of a zillion built-in functions in your formulas. The functions include random-number functions, for example **random( )**, which returns a random number between zero and one. So you might have a collection of 100 cases with an attribute called **T**. If you give **T** that formula, the 100 values will all become (different) random numbers.

Note: when you enter a formula, enter only the part of the equation to the right of the equals sign. For example, to enter the formula for a new attribute, **area**, use **length * width**, *not* **area = length * width**.

If you select the collection or the case table or a graph, you can then choose **Rerandomize** from the **Analyze** menu, and all of the **T**s will get new random numbers. All views will update.

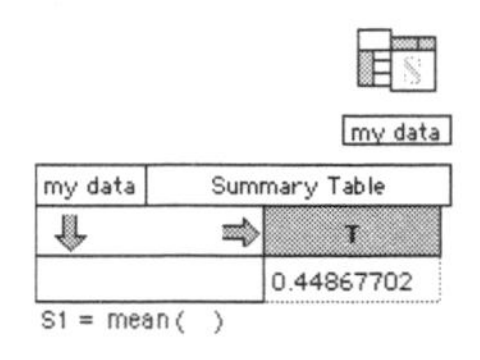

Sometimes, instead of looking at a graph, you just want to see the numbers. Then you need a *summary table*. As with graphs, you drag attributes to the table to display information about them. If it's a quantitative attribute, the summary table displays the mean by default. If you want a different statistic, select the table and choose **Add Formula** from the **Summary** menu. Use the formula editor to write a formula for whatever you want displayed.

A *measure* is an attribute that applies to the entire collection, for example, the mean of these random numbers. It would have its own name (**meanT**, say), defined by a formula (e.g., **mean(T)**) in the **Measures** panel of that collection's inspector.

Then (and here's the way cool part) you can automate this process, collecting means of these 100 values into a new collection where you can analyze them. This is called *collecting measures*. Select the original collection, then choose **Collect Measures** from the **Analyze** menu. Fathom creates a new collection and collects five measures by default.

To control how many measures you collect, use the inspector for the *measures* collection. The **Collect Measures** panel in the inspector lets you specify the number of measures, whether the collection empties when you collect, and so forth. You can read about collecting measures in detail; see "Collecting Measures" on page 16.

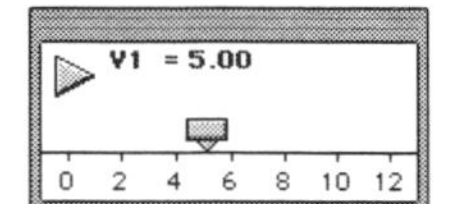

Finally, *sliders* are variable, global parameters. Drag one off the shelf to create a new one. You can change the name of a slider, and rescale it like any axis.

Ready to begin? Find one of the fifty demos in this book and give it a try. We start with a few demos about measures of center and spread...

Measures of Center and Spread

What does *mean* really mean? How can we get a feel for standard deviation? These are some of the basics, and without these topics, sampling distributions and confidence intervals won't make a lot of sense.

The demos in this section address some interesting properties of measures of center and spread, as well as giving you some of the background that a dynamic Fathom document can do so well.

Why pay attention to this kind of thing? Even if you're an expert in center-and-spread, you will see throughout the book that, especially with small samples, the *spread* of values plays an unexpectedly powerful role. We're used to the measures of *center* as the Most Relevant Summary—but it just isn't always the case.

In this section, you will find:

"The Meaning of Mean" on page 22. Here we see what happens to the mean when you drag a point. The mean moves, but how far?

"Mean and Median" on page 24. It's easy enough to say that the median is a resistant measure. But what does that really mean?

"What Do Normal Data Look Like?" on page 26. Here you look at some random numbers pulled from a normal distribution. You control the mean, standard deviation, and the size of the sample. The large sample shows the characteristic bell curve—but the small sample does not.

"Transforming the Mean and Standard Deviation" on page 28. When you add a constant to every data value, what happens? You change the mean, but not the standard deviation. What if you multiply by a constant? In this demo, you get to see it all happen dynamically; you'll be better able to "feel" the effect of these basic transformations after watching it happen.

"The Mean is Least Squares, Too" on page 30. This is a subtle, deep demo about the meaning of mean, maybe best approached when studying more advanced topics (such as the sample variance). Basically, the demo shows how you can define the mean as the number that minimizes the sum of the squares of the deviations from the data points. And what if you used absolute deviation instead of squares? Then you get the *median*.

web

Remember: If you see web in the side margin, that means that there's a solution on the web at the time of publication. See **http://www.eeps.com**.

Demo 1: The Meaning of Mean

The mean • How individual values affect the mean

This simple demo helps you get some “feel” for how the mean works. You could just drag points and see what happens. The instructions, below, take you down a particular path; follow it as far as you like—then explore!

What To Do

- Open the file **Meaning of Mean.ftm**. It should look like this:

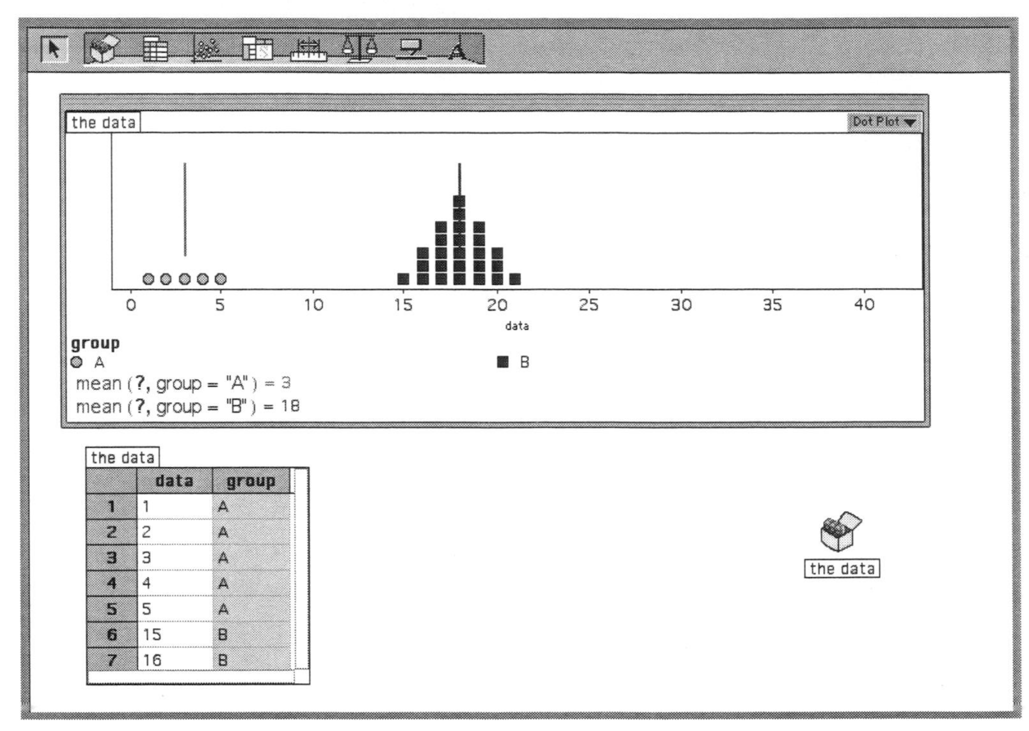

Note: To get big fat points like the ones in the illustration, choose **Preferences** from the **Edit** menu and check **Use Large Fonts**.

The most important thing here is the graph—a dot plot—showing the values of some data points. They’re in two groups, **A** and **B**—represented by circles and squares, respectively. The graph also displays the mean value of each separate group as vertical lines and as numbers, below. At the bottom, you can see a *case table* that shows the first seven values—all five in group **A** and the first two cases in group **B**.

- Grab the right-hand circle (at **data = 5**) and drag it. Notice that the mean changes and that its value changes in the table: look in the fifth row.
- Figure out how far you have to move that fifth point to have the mean grow by one, i.e., from 3 to 4. (If you can’t get it exactly by dragging, you can get close, and then type the exact value into the fifth row of the case table and press **enter**.)
- Put the point back at 5. (You can drag it there, type the value, or use **Undo** repeatedly.)

The shortcut for **Undo** is **clover-Z** (Mac) or **control-Z** (Windows).

- Drag a different point in group **A** and again, change the mean from 3 to 4.
- Now let’s look at group **B**. Begin with the lowest “square” point—the one at **data = 15**. How far do you have to move it to increase the **B** mean by one, from 18 to 19? (You may have to rescale the axis to move it that far. See “Rescaling Graph Axes” on page 18.)

- Again, put your point back where it started, at 15.
- Select several points in the **B** group by dragging a rectangle around them.
- Now drag the whole group (point at one of them and drag). See how far you have to move it to get the mean to change by one.

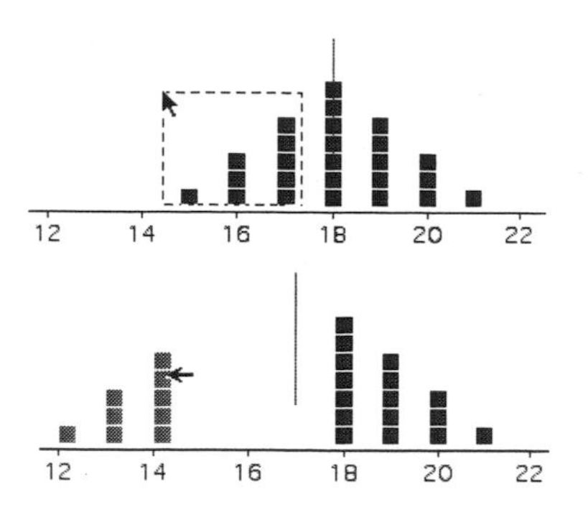

You should see that a point in a small data set has more influence on the mean than a point in a larger one. Similarly, moving more points moves the mean faster.

Note

This whole idea of "influence" requires both an actor (the moving point) and an act-ee (in this case, the mean). As you'll see in "Mean and Median" on page 24, outlying points have *no* influence over the median when they change a little. In "Least-Squares Linear Regression" on page 34, a moving point (the actor) acts on a linear regression line.

Sometimes students confuse "influential" with "outlier." In our case, every point has the same influence whether it's an outlier or not. With the regression line, a point right on the line—but far from the center of the cloud—has a lot of influence over the line's slope.

Extension

This extension will help you see the "balance point" meaning of the mean.

- Start with the file the way it started. You may need to re-**Open** it if you've done a lot.
- Click once on the graph to select it.
- Now choose **Plot Value...** from the **Graph** menu. The formula editor opens.
- Enter **mean( )**. (The empty parentheses are correct.) Press **OK** to close the editor. A new line and value appear—the mean of the whole data set, both groups together. Note its value. (It should be 15.5.)

You should be able to see this point as the balance point for the whole data set; imagine all of the points as blocks on a board, and the new line as the fulcrum. Let's explore it a little more:

To get these values to be *exactly* 3 or 18, just edit them in the case table.

- Stack all of the points from group **A** onto that group's mean (3). See if the group mean has changed.
- Stack all of the points from group **B** onto its mean (18). (To see all of the points, you will need to stretch the graph vertically.) Again, has the group mean changed?

Questions

If you are working through these demos on your own as a learner, these questions are for you, to help solidify your understanding. If you're a teacher using this as a demo for a class, these are suggestions for the kinds of questions you could ask your students.

1 How far is it from the **A** mean to the total mean? How far from the **B** mean to the total mean?

web

2 There are 5 points in group **A** and 25 in group **B**. How do those quantities relate to the distances from the mean?

Demo 2: Mean and Median

Measures of center: mean, median, and midrange • Resistance: what happens to the measures when you move one point

I've known how to find the mean and the median of a set of numbers for ages, but I never *really* understood what was important about the difference between them until I played with something like this.

This demo explores the properties of these two measures of center—the mean and the median. In particular, it's about how they are or are not resistant to changes in the data. It's important not to be judgmental here; being resistant is not good or bad—but the resistance of the measure is one factor that helps you figure out which one you want to use. The point is to come up with a number that represents the entire data set fairly. You might think of this as a "typical" value.

As with "The Meaning of Mean" on page 22, the main idea here is to drag points and see what happens.

What To Do

- Open the file **Mean and Median.ftm**. It should look like this:

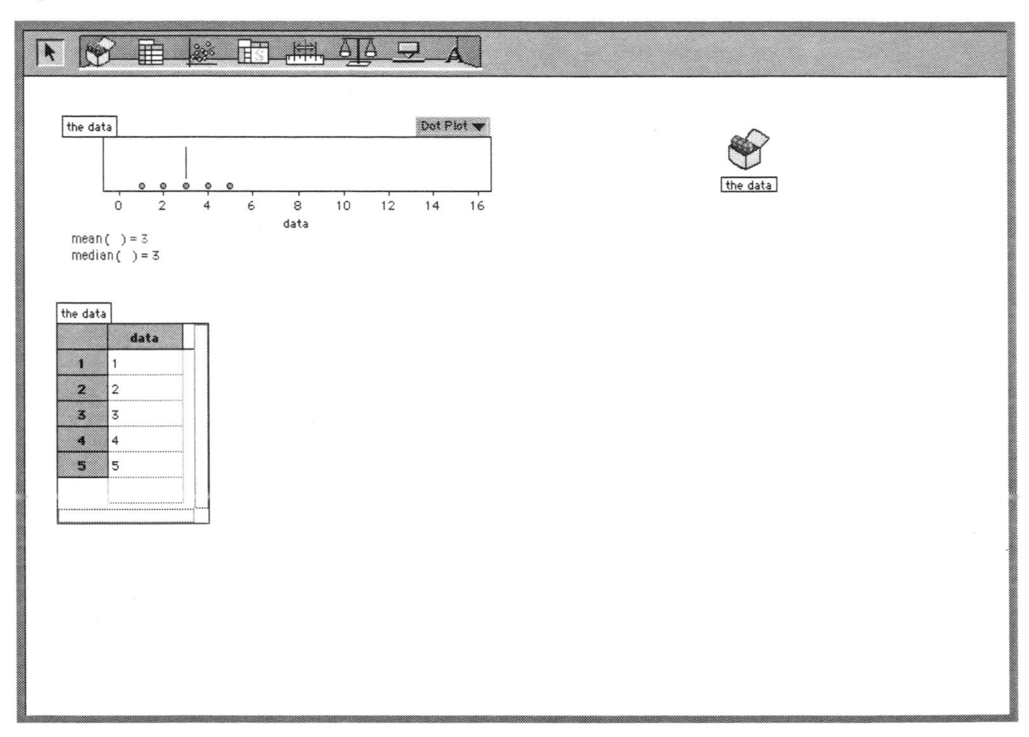

Here you see a graph showing **data** values and, below it, a case table. The lines for the mean and median are in the same place (at **data** = 3) right now.

Predict what will happen before you move the point!

- Grab the right-hand point with the mouse and move it to the right. See how the line for the mean moves and the median does not. Also see how the numerical value for the mean (just below the graph) changes and how the data value for the point changes in the table.
- Move the right-hand point (originally the one with value 5) to the left, past the median. Watch how (and for how long) the median sticks to the point you're moving.

Questions

1. How far do you have to move the right-hand point to get the mean to increase by one (i.e., from 3 to 4)?
2. Why doesn't the median change when you wiggle the point off to the right?
3. Why does the median stick to the moving point when you move it past the median?

web

4 Over what range of values does the median stick to the point you're dragging?

5 How is all of that different (or the same) if you drag a different point?

Harder Stuff

6 Explain (in a few sentences or a short paragraph; or in a lively discussion) why you have to move the point as far as you do to get the mean to change by one.

7 Explain why it doesn't make a difference which point you move.

8 Statisticians say that the median is "resistant to changes in outliers" while the mean is not. Discuss whether *resistant* is a good word to use for this.

9 Aloysius says, "the median is a better measure of center because it doesn't go all wacko if you get one crazy point." Hildegarde says, "the mean is better because it takes all the data into account; the median really only depends on one or two points." Discuss the ways in which both are right.

10 Give examples where you would prefer to use the mean to determine a "typical" value for a data set; give other examples where you would prefer the median.

More To Do

The shortcut for **Undo** is **clover-Z** (Mac) or **control-Z** (Windows).

- Put the collection back the way it was. (Re-**Open** it or choose **Undo** from the **Edit** menu repeatedly until you get back to the original state.)
- In the case table, click in the empty box at the bottom of the **data** column. Type **6** and press **enter** or **return**. You have just created a new case (and a new point on the graph; you may have to change the scale to see it).
- Now drag the (new) right-hand point. What's different when you drag to the right? What's different when you drag to the left?

Extension: Plotting the Midrange

Mean and median are not the only measures of center. Another one is called the *midrange*. Let's plot it and see how it behaves.

- Put the collection back so you have five or six points, values {1, 2, 3, 4, 5, and maybe 6}.
- Click on the graph to select it.
- Choose **Plot Value** from the **Graph** menu. The formula editor opens.
- Enter this formula: **(min( ) + max( )) / 2**, as shown. (You can type those characters exactly; notice how typing the ")" works.)

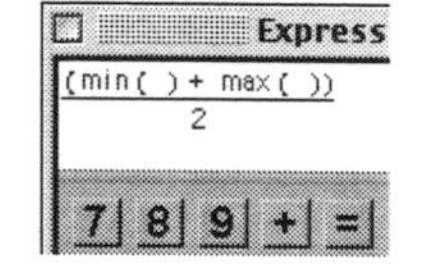

- Close the formula editor by pressing **OK**.
- Drag a point around. See what happens.

Questions about the Midrange

11 How far do you have to move the right-hand point to increase the midrange by one?

web

12 Explain why it's that far.

13 Over what range of values for the maximum point (i.e., the one on the right—with a value of 5 or 6) does the midrange not change? Explain why.

web

14 How would you characterize how "resistant" the midrange is to changes in outliers—especially compared with the mean and median?

15 What are advantages and disadvantages of using the midrange as a measure of center?

Demo 3: What Do Normal Data Look Like?

Normally-distributed data • The effect of changing the mean and standard deviation

A lot of things in traditional statistics depend on data being roughly normal. But what do such data look like? We may have a mental image of the ideal bell curve, but a real sample from a normally-distributed population may look very different.

What To Do

- Open the file **Normal Data.ftm**. It will look something like this:

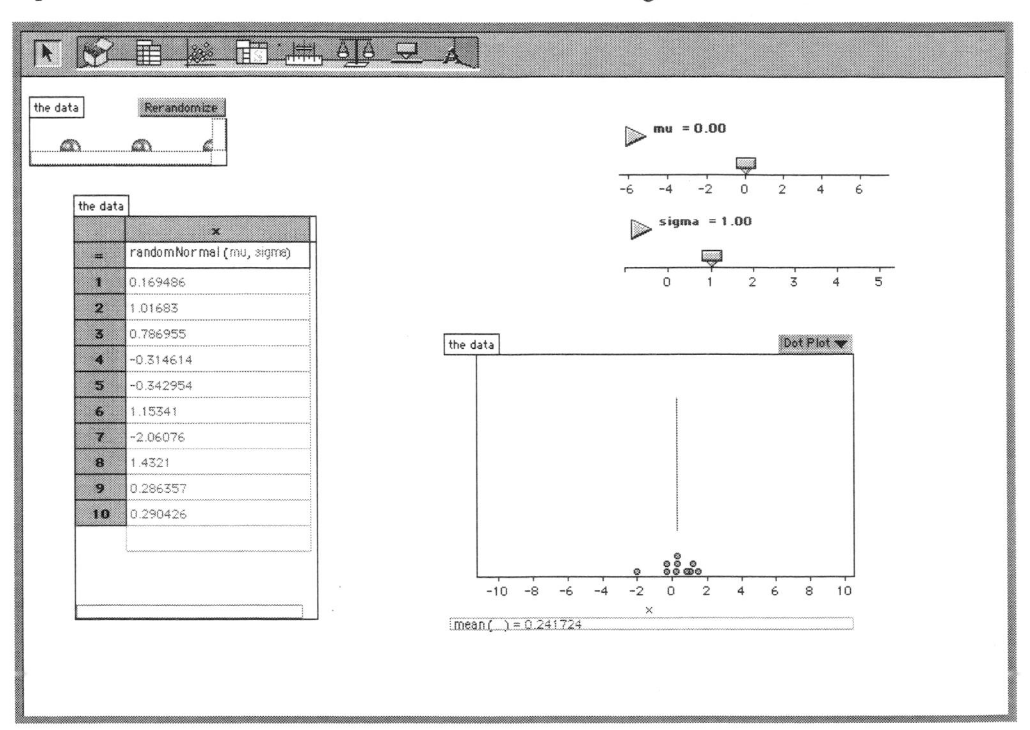

In the upper left is the collection (the box of balls), opened to show its **Rerandomize** button. Two *sliders*, **mu** and **sigma**, control the mean and standard deviation of the population, respectively. The case table and the graph show the data.

- Press the **Rerandomize** button repeatedly to see how the data change. Observe how the mean of this sample of 10 numbers jumps around.
- Change the mean by dragging the **mu** slider around.

The data also rerandomize whenever you move the sliders.

- Change the standard deviation by changing **sigma**.
- Change the dot plot to a histogram by choosing **Histogram** from the pop-up menu in the graph.
- Again change **mu** and **sigma** (you may want to rescale axes; see "Rescaling Graph Axes" on page 18) to see how this looks.
- Change the graph to a **normal quantile plot** (use the pop-up menu). In this plot, if the data are perfectly normal, they will lie on a straight line.
- Rerandomize the data repeatedly to get an idea of the variety of "lines" you can get with a sample of 10 points drawn from a genuine normal distribution.

Questions

For these questions, display the data as a histogram or a dot plot:

1. With **mu = 0** and **sigma = 1** (this is called the *standard normal distribution*), what's a typical range of values for the points in your sample?
2. What's a typical range of *means* for your sample of 10 from a standard normal distribution?
3. Set **mu = 1**. Now what's a typical range of values?
4. Set **mu = 0** and **sigma = 2**. Again—what's a typical range of values?
5. About what proportion of the time do the data look fairly symmetrical with a hump in the middle?

You should see that with a sample of 10, the graphs often look far from normal—they often don't have a hump in the middle, and they aren't symmetrical. The normal quantile plots can be pretty far from straight lines. So let's add some more points.

Onward!

- Click on the collection (or case table or graph) once to select it. Then choose **New Cases...** from the **Data** menu. Enter 190 and press **OK**. Now you have 200 cases in your distribution.
- Play with the displays and sliders as before. Observe how different it is from the sample of 10.
- Let's put the curve on the graph. First, make the graph a histogram with the pop-up menu.

With a *density scale*, the vertical axis shows the proportion that are in the bin *per unit on the x-axis*. This way, the total area under the curve is 1.

- Choose **Scale>Density** from the **Graph** menu. The vertical axis will change to a density scale.
- Choose **Plot Function** from the **Graph** menu. The formula editor will open.
- Enter **NormalDensity(x, mu, sigma)**. Press **OK** to close the editor. You should see something like the graph in the illustration.
- Again, play with the sliders to see how the graph changes.

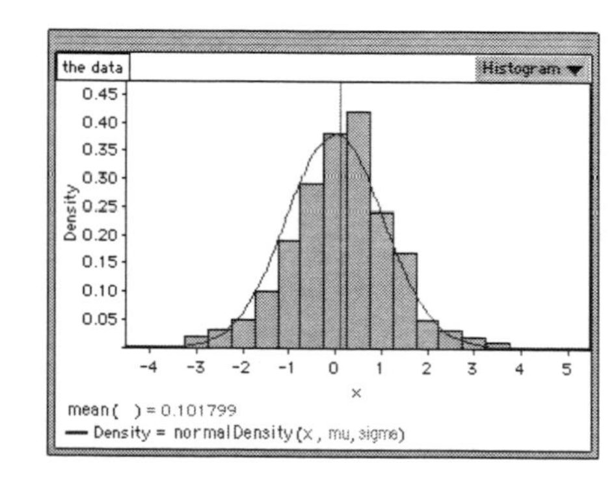

More Questions

web

6. With 200 cases, about what proportion of the time do the data look fairly symmetrical with a hump in the middle?
7. In the normal quantile plots, (choose **Normal Quantile Plot** from the pop-up menu in the graph) where do the points most deviate from the straight line?

Extension

Finally, let's see how this normal curve looks with fewer cases.

- Drag a rectangle to select most of the bars in the graph. They will turn red.
- Choose **Delete Cases** from the **Edit** menu. Those bars will vanish.
- Drag the sliders or rerandomize to see how the normal curve looks with less data.

Demo 4: Transforming the Mean and Standard Deviation

What happens to mean and standard deviation when you add a constant to every value or multiply every value by a constant

One issue that deserves a dynamic demo is what happens to the mean and standard deviation of a set of data when you add a constant to every data value or multiply the data values by a constant.

What To Do

- Open the file **Transform Mean and SD.ftm**. It should look like this:

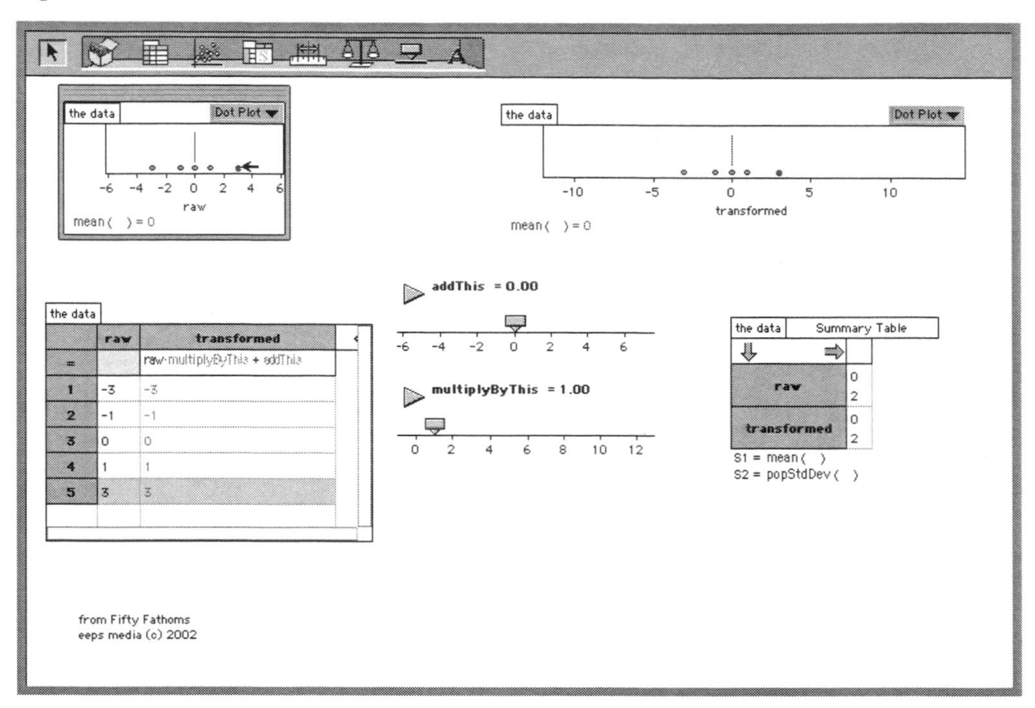

Note that the **raw** and **transformed** axes are not linked; if you rescale one, the other will not change.

The main thing to note is that there are two graphs—one of the **raw** data, and one of the **transformed** data. Also note that the original data—{−3, −1, 0, 1, 3}—has a standard deviation of 2, as you can see in the *summary table* bottom right. If you've never seen a summary table, take a moment to figure it out.

- Drag the highest point in the **raw data** graph (with **raw = 3**—where the arrow is in the illustration) and see how it changes the mean and the standard deviation in the summary table. You can also see the mean change on the graph itself. Note that the transformed data mirrors the raw data exactly.

Undo is in the **Edit** menu. Its shortcut is **clover-Z** (Mac) or **control-Z** (Windows).

- In the case table, re-type "**3**" as the highest data value and press **enter**. The display should now be as it was when you began—with a mean of 0 and a standard deviation of 2 for both the raw and transformed data. (If necessary, simply close and re-open the file.)
- Play with the slider called **addThis**. Note what happens to the transformed data.
- Also note what happens to the mean and standard deviation of the transformed data.
- Return **addThis** to zero. (Use **Undo**, or just edit the number in the slider.)
- Play with the slider called **multiplyByThis**. Watch what happens. When you're done, return its value to 1.

Note that you can observe these phenomena quantitatively by comparing the values on the sliders to the values in the summary table. That is, when you change **multiplyByThis** from 1 to 2, how does the standard deviation change?

Questions

1 What happens to the mean if you add x to every value in a data set?

2 What happens to the standard deviation?

3 What happens to the mean if you multiply every number in a data set by x?

4 What happens to the standard deviation?

Harder Stuff

The two questions that follow are not really about transformations of the entire data set, but rather about the effect of dragging a single point on the standard deviation. You could think of these as an extension of "The Meaning of Mean" on page 22 and "Mean and Median" on page 24.

5 In "Mean and Median" on page 24, you dragged the point on the far right to see how far you had to drag it in order to change the mean by one. (You had to drag it five units.) Check that, and see how far you have to drag that point to increase the *standard deviation* by one.

web

6 Do the same for the *middle* point in the distribution. How far do you have to move it to increase the mean by one? How far to increase the standard deviation by one? Compare that to moving the extreme point and explain why it's the same or different.

Demo 5: The Mean is Least Squares, Too

Defining the mean as the place where the sum of squares of deviations is a minimum (just like the least-squares line) • The median, and what it minimizes

We have put this demo here in the opening section because it is about the mean, and measures of center come early in this book. But this is a sophisticated demo. It may be better to think of this as following "Least-Squares Linear Regression" on page 34—which is more complex on the surface—with this subtle twist on an old friend.

What To Do

- Open the file **Mean Is Least Squares.ftm**. It will look like this:

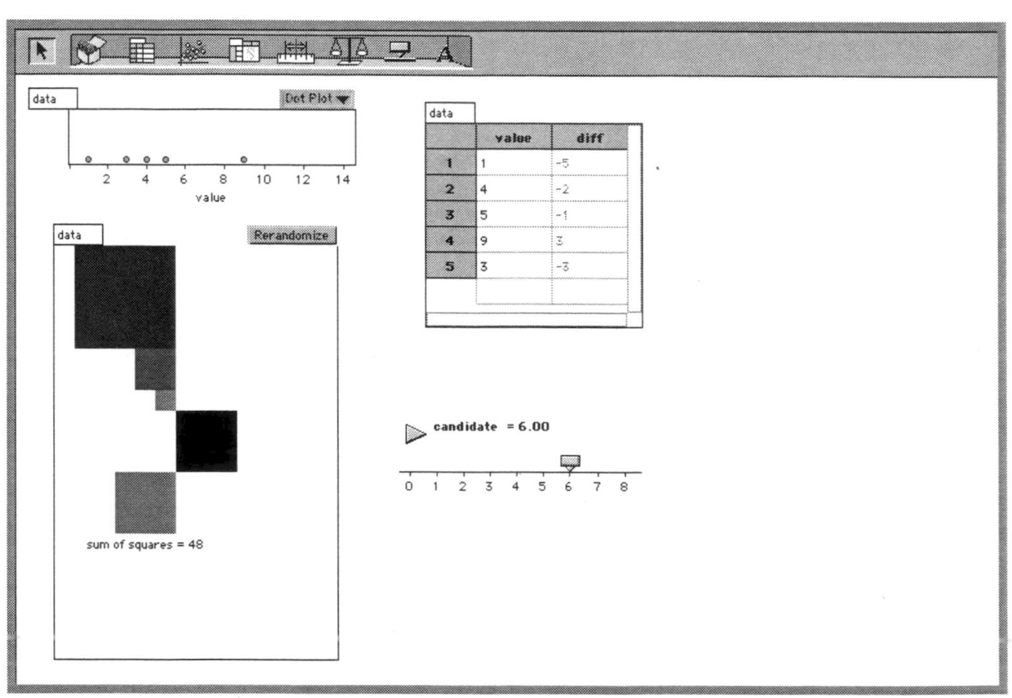

Note: the **Rerandomize** button won't do anything because there are no random values.

Here we have five data points and a slider in the middle called **candidate**. That slider controls a number that we are proposing for a measure of center. We're essentially asking, "how good is this number—6.0 at the beginning—as a measure of center for these five data points?" (Lousy. But we're getting ahead of ourselves…)

The collection itself (usually gold balls) this time shows colored squares. One vertical edge of each square represents the data value; the other is at the candidate value. You can also see that we have computed a quantity called **diff** for every data point, which is the difference between that point and the candidate (equal to the length of the side of each square).

- Drag data points in the dot plot (upper left) and see how the values—and the squares—change. Notice how the position of the dot corresponds to the vertical edge of one of the squares. Also, note how the sum of squares changes.
- Drag the slider, **candidate**, and see how things change. Try to move it so that the sum of squares (at the bottom of the stack of boxes) is a minimum.

It would be lovely to have Fathom collect all those sums of squares. Inside the **data** collection, we have computed two measures. One, called **proposedCenter**, simply reports the value in the slider. But the other, **sum_squares**, is the sum of the squares of **diff**. That's the thing we're interested in looking at for different values of **candidate**.

It's confusing that **proposedCenter** has a different name from **candidate**. But Fathom needs to distinguish between the attribute values and the slider.

- Open the file **Mean Is Least Squares 2.ftm**. It will look like the previous one except that it has a measures collection, and a graph showing how **sum_squares** depends on **proposedCenter**. We've already collected one value—the sum of squares is 48 when **proposedCenter** is 6.
- Drag the **candidate** slider. New points will appear on the measures graph, and the picture of the squares will update. Keep dragging until you have a clear minimum for **sum_squares** in the graph. The graph will look something like the one below left.
- Holding down **option** (Mac) or **control** (Windows), drag a rectangle around the minimum (shown). When you let go, Fathom will zoom to that rectangle, and you'll see a clearer minimum, as at right below.

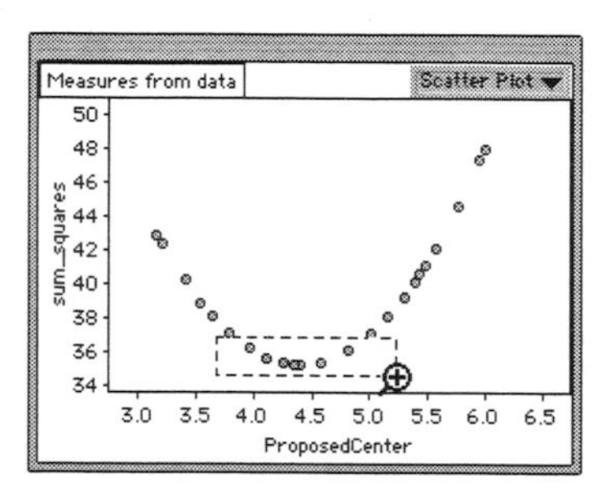

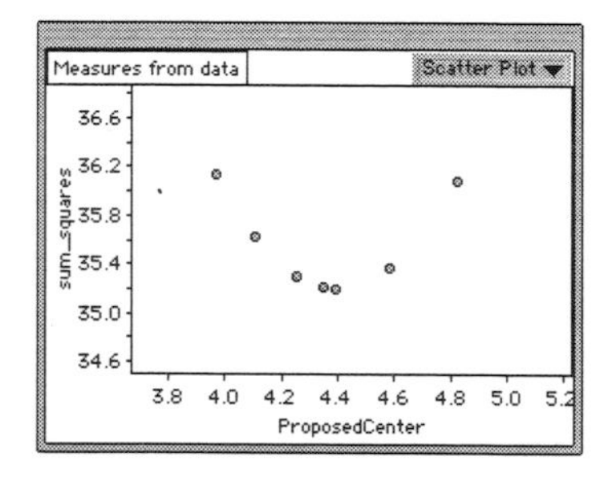

- Add new points (by moving the slider to the relevant area) as you see fit. You may even want to zoom into the slider to get finer values for **candidate**. Determine the value for the candidate that gives the smallest sum of squares of the differences.

This should be the mean, in this case, 4.4, and a sum-of-squares of a little over 35. That is, the mean minimizes the sum of squares of the differences—*the mean is a least-squares statistic.*

How to delete the cases in the measures collection.

- Change some of the data values so the mean is different.
- Delete all of the cases in the *measures* collection (not in the data collection!). One easy way is: select the graph; choose **Select All Cases** from the **Edit** menu; then choose **Delete Cases** from the **Edit** menu.
- Drag the slider again to produce the graph of sums of squares. See if that minimum matches the new mean.

Extension—What If We Don't Use Squares?

Adding squares of differences is a great way to get a reasonable measure of how good a candidate measure of center is. But it's not the only way to do it. In fact, when we ask people to devise their own measure for how good a center is, one of the most popular is to add up the absolute distances of the points from the center. Speaking graphically, add up the lengths of the line segments connecting the points to the center instead of the areas of the squares. Speaking symbolically, minimize

$$M = \sum |x-c| \text{ instead of } M = \sum (x-c)^2$$

where c is the candidate center.

Actually, to facilitate this extension, you've been collecting sums of absolute differences all along. All we need to do is graph them:

- Double-click the measures collection (the box of gold balls) to open its inspector.
- Click the **Cases** tab to bring up the **Cases** panel.
- Drag the name **sum_abs_diff** (it will probably appear as **sum_abs_di...**) from the inspector (shown) to the vertical axis of the graph, replacing **sum_squares**.
- Drag the slider as necessary to fill in the graph. Where is the minimum?

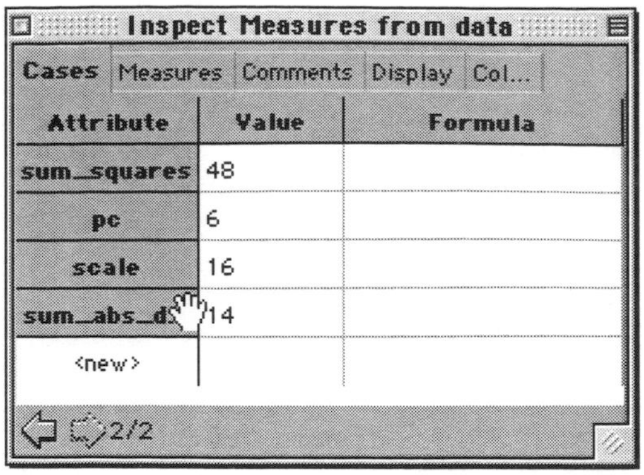

Inspect Measures from data

Cases | Measures | Comments | Display | Col...

Attribute	Value	Formula
sum_squares	48	
pc	6	
scale	16	
sum_abs_d	14	
<new>		

2/2

Extension Questions

web 1 What special value minimizes the sum of absolute differences?

2 The stack of colored boxes in the collection on the left is shortest when this number is a minimum (not when the sum of squares is a minimum). Explain why.

3 Remove a case from the original collection: select one of the colored boxes in the stack and choose **Delete Case** from the **Edit** menu. Now there are four instead of five. Now redo the graph of the measures (delete all cases in the measures collection, as described on page 31, and play with the slider). What do you notice about the graph? What's the minimum value? Explain.

web 4 Suppose that instead of adding up squares or absolute differences, we minimized the (absolute) difference between the candidate and the *farthest* data point. What would the graph look like? What value would you get for a minimum?

5 If you minimize the sum of absolute distances, or if you minimize the farthest distance (as in the previous question), your function of the candidate position is piecewise linear, unlike the smooth parabola we got for the sum of squares. What difference, if any, does that make in how useful these various minima are as measures of center?

Regression and Correlation

This section is all about the relationship between two continuous variables. When you use one to predict the other, you can use—among other things—a least-squares regression line. When you want to describe how two variables are related, you can use the correlation coefficient. Regression lines often show up in the "modeling" section of a stats course, and they belong there. But these techniques permeate inferential statistics too, as we shall see.

These topics can be tricky; the demos in this section will help clarify a few of the most troubling issues.

In this section, you will find:

"Least-Squares Linear Regression" on page 34. What are the *squares* in least squares regression? In Fathom, you can show them in this simple but essential demo.

"Standard Scores" on page 36. Just as we use proportions to compare two groups of different sizes, we can use standard scores to compare—or, in this case, combine—data from different distributions. The standard score is a measure of where something is in relation to the population, as measured in terms of standard deviation. This "dimensionlessness" is crucial to its success—and we'll use the same principle to create the t statistic a little later.

"Devising the Correlation Coefficient" on page 38. Bit by bit, we construct a statistic that measures the strength of a relationship, using standard scores as a starting point. The elegance of cross-multiplying may inspire you to be creative in the way you design your own statistics.

"Correlation Coefficients of Samples" on page 41. In a way, this is skipping ahead, giving you an early look at another big theme in these demos: sampling distributions. We will see that a sample may have a very different correlation than its source population. Things can seem correlated that aren't, and conversely.

"Regression Towards the Mean" on page 44. Why is it that the people who do best seem to fall the farthest? Fathom can show you. This phenomenon also helps us see how the regression line and the correlation coefficient really measure different things.

web

Remember: If you see web in the side margin, that means that there's a solution on the web at the time of publication. See **http://www.eeps.com**.

Demo 6: Least-Squares Linear Regression

Explore the squares in least squares • Minimizing the areas of the squares built on residuals

This is a must-see for anyone learning about statistics. As far as we know, this was first created by Bill Finzer using *The Geometer's Sketchpad*. So when Bill made Fathom, he included it in the guts of the program.

This demo clarifies what the least-squares line means by connecting the idea behind the calculation with a dynamic visual representation. It's so easy and quick, we've added a little more: the extension (below) explores the startling effect of influential points.

What To Do

- Open the file **Least Squares.ftm**. It shows made-up "data" that's pretty well correlated. It should look something like this:

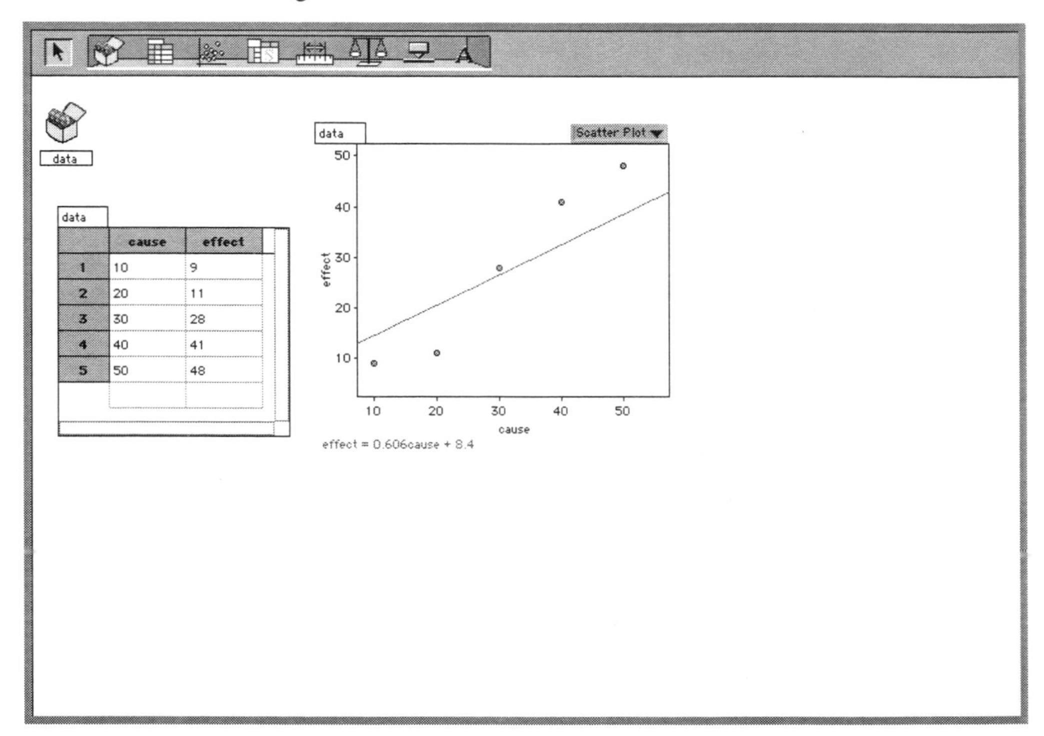

- Play with the brown *movable line*. Drag its middle to change the intercept; drag the ends to change the slope. Notice how the equation for the line (below the graph) changes as you move the line, and that the variables in the equation are the attributes **cause** and **effect** rather than x and y.
- Move the line to make as good an "eyeball" fit to the points as you can.

There may be several lines that appear to be "best" fits. In fact, different people will choose different best lines. So the question is, is there an objective way to determine how good a fit is? We'll explore one such way—looking at the squares of the residuals.

If there is no **Graph** menu, click once on the graph to select it. Then the **Graph** menu appears.

- Choose **Show Squares** from the **Graph** menu. Squares appear, built on vertical segments extending from each point to the line. In addition, you can see the **Sum of squares** down by the equation. It's the sum of the *areas* of the five squares.

If this is a whole-class demo, you can invite students to the computer to move the line.

- Play with the line, watching as the sum of squares changes. Try to minimize it. It will end up looking something like the illustration (but you can do better).
- Test your minimum; have Fathom display the true *least squares line*. Choose **Least-Squares Line** from the **Graph** menu; it appears with your movable line.
- You can try your hand at minimizing again if you wish: un-choose **Least-Squares Line** to make the answer go away, then move any points you want by dragging them. Now minimize the sum of squares again by dragging the movable line, and test your result by making the least-squares line reappear.

Questions

1. How did you go about minimizing the sum of squares?
2. It should look as if some points contribute more to the sum of squares than others. Which ones?
3. Why do you suppose they used *squares* for this? web

Extension

- Make the least-squares line appear and make the movable line disappear (un-check **Movable Line** in the **Graph** menu)—that is, just show the least-squares line and its associated squares.
- Drag a point and watch what happens. Then return the points to their original places.
- Drag an end point up and down a little. What happens to the least-squares slope? What happens to the intercept? Does it matter which end point you move?
- Drag the middle point up and down a little. What happens to the slope? What happens to the intercept?
- Now we want to drag points farther, so we need a bigger range on both axes. Rescale the axes (See "Rescaling Graph Axes" on page 18.) to give more space, as in this picture:

To zoom out automatically, hold down **option-shift** (Mac) or **control-shift** (Windows) and point at the graph. A magnifying glass with a minus-sign appears. Click to zoom out.

- Now drag a point and watch the line move (its equation updates, too).
- See if you can make the least-squares slope *negative*.
- See if you can make **r^2** (its value appears at the bottom of the graph, to the right of the equation) equal to zero. What do you suppose that means?

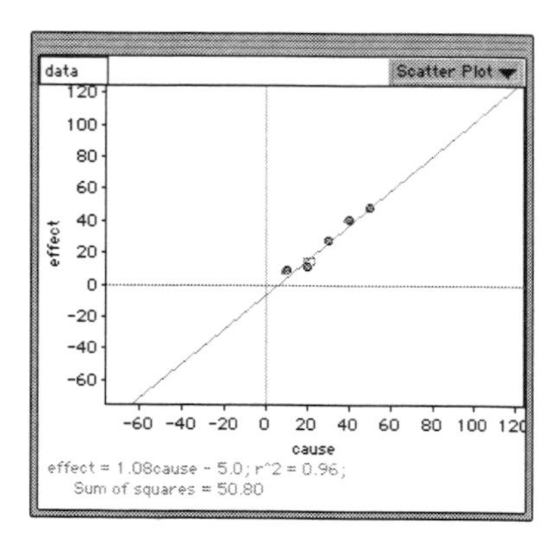

Demo 7: Standard Scores

Using standard scores to compare unlike scales • Making a scale in terms of standard deviations

Suppose we want to come up with a fair way of combining scores on two tests. Unfortunately, the scores are very different—they have different centers and spreads. The first score, like the SAT, can range only between 200 and 800; unlike the SAT, its scores pretty much blanket that range. The second score can take any value between 0 and 100, though people generally scored above 70. Furthermore, since these tests are in completely different areas, they are uncorrelated. Nevertheless, we have to come up with a combined score. (Think of the biathlon in the Winter Olympics, where the contestants ski and shoot; there has to be some way of awarding the medals.)

Let's begin by trying the simplest solution—just adding the two scores together. We will see why that's a bad idea.

What To Do

- Open the file **Standard Scores.ftm**. It looks something like this:

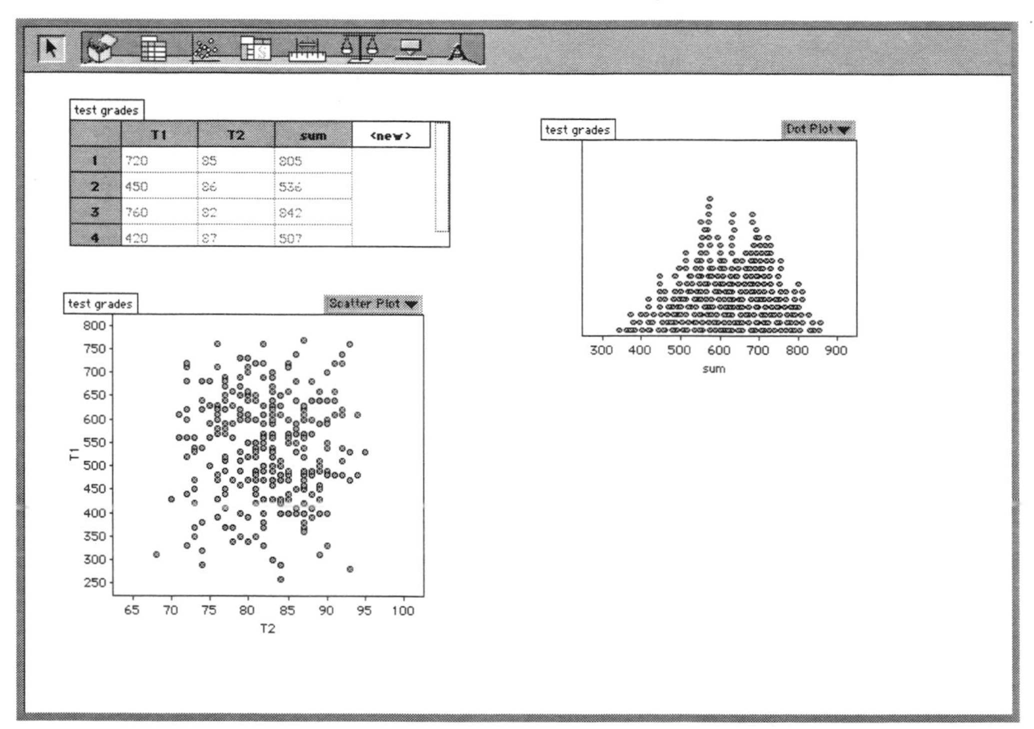

Here you can see the top of a table of test scores **T1** and **T2**, and their **sum**. The lower left graph shows **T1** plotted against **T2**; the one in the upper right shows the distribution of their sum.

Let's see why simply adding is such a lousy way to combine the scores in this situation:

- With your mouse, drag a rectangle around the highest couple dozen scores in the **sum** graph, and release. The points you have selected turn red—and so do the *same* cases in the other graph.

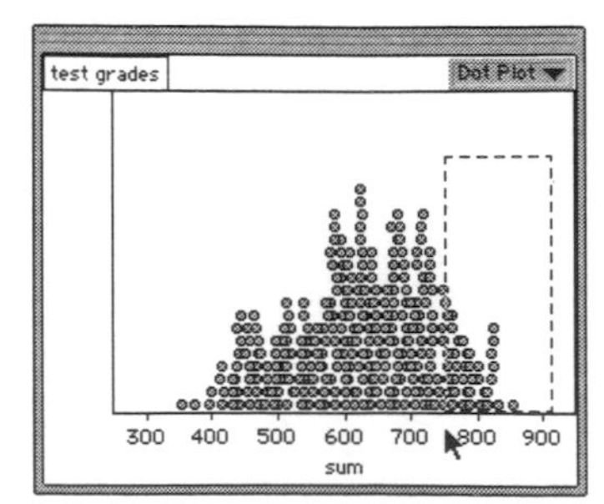

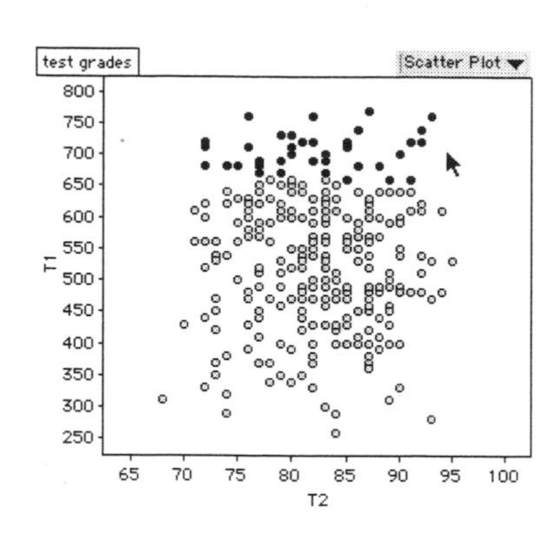

You should see a graph like the one in the illustration. The band of selected points is essentially horizontal. Those points all have a high score on **T1**, *whatever their performance on **T2***. That is, if we just use the sum, it's pretty much the same as ignoring **T2** entirely!

The problem is that the scales of the two scores are so different. Just because the numbers are high for **T1** (and have such a range) doesn't mean it should dominate. One strategy is to transform the scores so they have the same center and spread—and then add. This is called changing the data to *standard scores*. Let's try it:

The shortcut for **Edit Formula** is **clover-E** on the Mac or **control-E** in Windows.

You can enter the formula many ways; one is to type those keys exactly (including all parentheses).

- In the case table, click in the **<new>** box at the top of the blank column. Name the new attribute **Z1** and press **return** or **enter**. (*Z* is a traditional letter to use for a standard score.)
- Select the new attribute by clicking once, and choose **Edit Formula...** from the **Edit** menu. The formula editor appears.
- Enter this formula: **(T1 – mean(T1))/s(T1)**. That is, take the difference of the score from its mean, and then divide the result by its standard deviation. That way, if your score on **T1** is one standard deviation above the mean in the group, your *standard score* for that test (**Z1**) will be +1.0. The formula will look like the illustration in the editor; when you have it right, close the editor with **OK**.

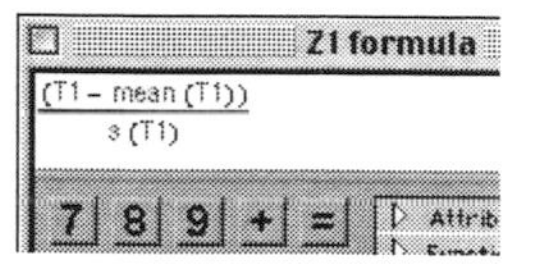

- Make another new attribute, **Z2**, analogously. You may have to scroll the case table to the right to get another **<new>** column; the formula is **(T2 – mean(T2))/s(T2)**.
- Make a new graph by dragging one off the shelf. Put it in the lower right where you have some space.
- Drag **Z1** to the vertical axis of the new graph and **Z2** to the horizontal axis. Note how the graph looks just like the **T1-T2** graph, only with different scales.
- Finally, make a third new attribute: **Zsum**. Give it the formula **Z1 + Z2**. Drag its name to the horizontal axis of the top graph, replacing **sum**.
- Now select the top couple dozen points in **Zsum**. Aha! Now the **T1-T2** graph shows the top combined scores as a triangular region in the upper right—the ones who did well on both, as you can see in the illustration.

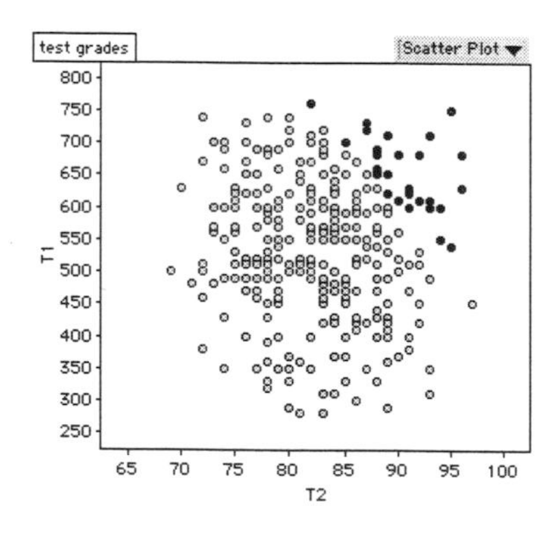

Note: A *z*-score is *dimensionless*, similar to the way *t* is, as we will discuss in "The Road to Student's t" on page 71.

Extensions

- For a good time, drag **sum**, and then **Zsum**, to the *middle* of either scatter plot.
- Edit the formulas for **T1** and **T2** to give them a differently-shaped distribution (e.g., exponential). Then use this machinery to see if you think this way of combining scores is fair for distributions that are not roughly bell-shaped.

Demo 8: Devising the Correlation Coefficient

How the correlation coefficient measures what it does

Many students learn that a correlation coefficient near +1 or –1 means a perfect correlation, and they may even do wonderful exercises where they learn to recognize what a correlation of 0.8 looks like. But this demo is about how the correlation coefficient *works*; in particular, how it is that the product of the two coordinates—suitably transformed into standard scores—helps measure what we're after.

What To Do

- Open the file **Correlation.ftm**. It should look like this:

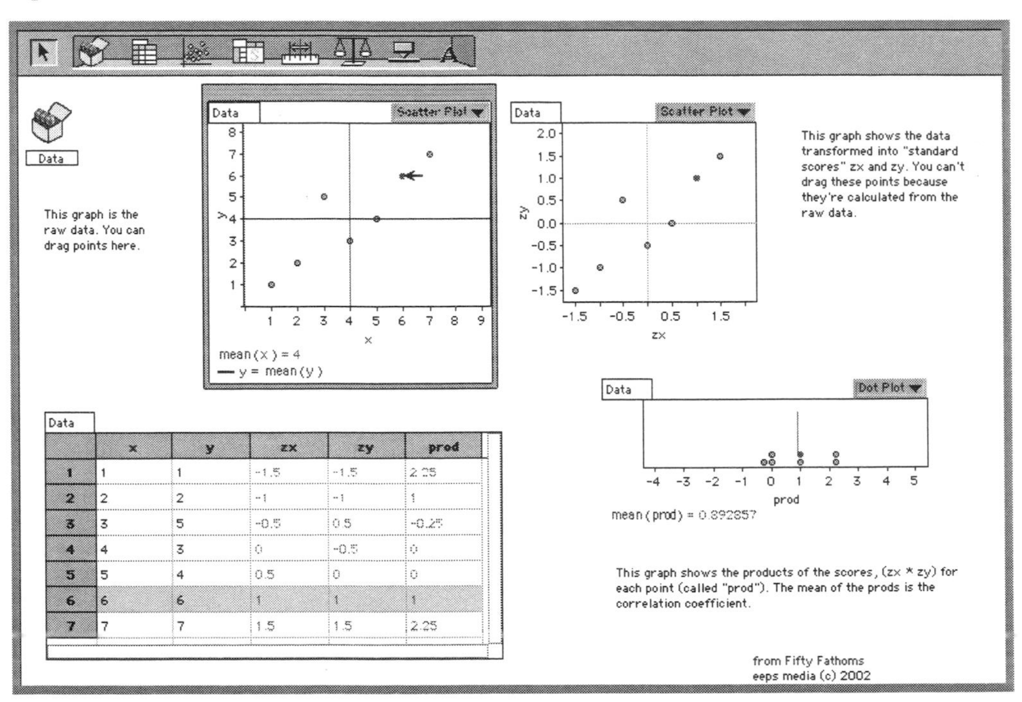

	x	y	zx	zy	prod
1	1	1	-1.5	-1.5	2.25
2	2	2	-1	-1	1
3	3	5	-0.5	0.5	-0.25
4	4	3	0	-0.5	0
5	5	4	0.5	0	0
6	6	6	1	1	1
7	7	7	1.5	1.5	2.25

The graph in the upper left—where the pointer is in the illustration—shows the raw data, with lines showing the mean of **x** and the mean of **y**. To its right is a similar scatter plot that shows the "standard scores" **zx** and **zy**. Below that graph is a dot plot showing, for each point, the product of the two coordinates. That is, it shows **prod** where **prod = zx * zy**. The graph also shows the mean of **prod**—which is the correlation coefficient.

- Grab a point (such as the one indicated in the illustration) and drag it around the left-hand graph, watching how that affects the right-hand graph. (Ignore the bottom graph for a bit.)
- Drop that point near the intersection of the two lines—the intersection of the means of the two variables. Watch how the intersection moves as you approach.
- Move each of the points to a place near that intersection, watching how the mean moves and how the points behave in the other graph. Close is good enough. *Save one of the most extreme points for last.*
- Finally, move the extreme point in to the intersection of means—and watch what happens in the other graph.

You should see that the standard scores—the *z*'s—mirror the data almost exactly, with a couple of crucial differences.

Questions

web 1 When you move a point, the others (on the **zx-zy** graph) first move in the opposite direction. Why?

web 2 As you move the point even farther in one direction, what happens to the other points?

3 How would you characterize the difference in shape between the top two graphs (that is, they look kind of the same, but when you drag a point far to the right, they don't look as similar)? How do you account for that difference?

4 When you brought that last point in to the mean, what happened? Explain why.

Onward!

If you are leading a demo, this is a great place to involve the class in deciding where to move points, etc.

- Return the points to their original positions (or close to them) though repeated **Undo**, re-**Open**ing the file, etc.
- Now, move points around again, but this time look at the image of the moving point in the **prod** (lower-right) graph. Note where the point must be for its **prod** to be negative.
- Move points so all the points have negative **prod**s. Notice where the points are.
- Now move the points so that they all have positive **prod**s. Again, notice the pattern.
- Make all the points line up perfectly with a positive slope. Notice the mean of the **prod**s (in the graph).
- Pick one point and move it perpendicular to the line you have made. Observe what happens to the positions of other points on the **zx-zy** graph, its **prod**, and the mean of the **prod**s. Then return the point to its place in line.
- Again move the points in towards the center, one at a time. Watch the position of the extreme point—the one you're going to move last—in the **prod** graph.

More Questions

5 Where does a point have to be to have a positive **prod**? A negative **prod**?

6 When the points are all lined up, they still don't have the same **prod**s. Which points have big **prod**s? Which have little ones?

7 If you line up all the points perfectly with a *steep* positive slope, the original graph looks very different from the way it does if you line all the points up perfectly with a *shallow* positive slope. How do their **zx-zy** graphs differ? Why?

8 Why did the **prod** for the extreme point get more extreme as you dragged the other points to the center?

web 9 How did the correlation coefficient—the mean of the **prod**s—change as you dragged the points to the center? Why?

Extension

- Move the points so that the mean of the **prod**s is zero. Do this in at least three radically different ways.

The Point

The correlation coefficient works because of the interplay of two big ideas.

- Looking at the products of the *x*'s and *y*'s is flat-out brilliant, because points in quadrants I and III contribute to the positive and II and IV to the negative parts of the correlation. Because the distances are all relative to the mean in each dimension, the sum of these products is a measure of how well lined-up all the points are.

- Standard scores—the *z*'s—scale the dimensions to show their spreads and distributions comparably, in a way that makes the original units and magnitudes irrelevant. Without using standard scores, you could still make numbers that measure how well lined-up points are, but you would not be able to compare those numbers to measures from other data sets.

Harder Stuff

10 When you line the points up, you can get a correlation coefficient—a mean of the **prod**s—equal to one. Why can't you get the mean to be *larger* than one? After all, some of the **prod**s are larger than one; why can't you arrange it so that more of them are? Try to make a correlation coefficient larger than one, and figure out qualitatively what happens to prevent you from succeeding. If you can, prove it with symbols.

11 Invent a correlation coefficient that's based on the median and IQR instead of on the mean and standard deviation. What possible values does it have? What are some advantages and disadvantages of using this measure of correlation?

Demo 9: Correlation Coefficients of Samples

How samples from a correlated population yield different values for the correlation • How sample size affects that sampling distribution

Note: This activity was inspired by *Activity-Based Statistics*, "Relating to Correlation" (first edition Instructor Resources, page 330).

What To Do

- Open the file **Correlation in Samples.ftm**. It will look something like this:

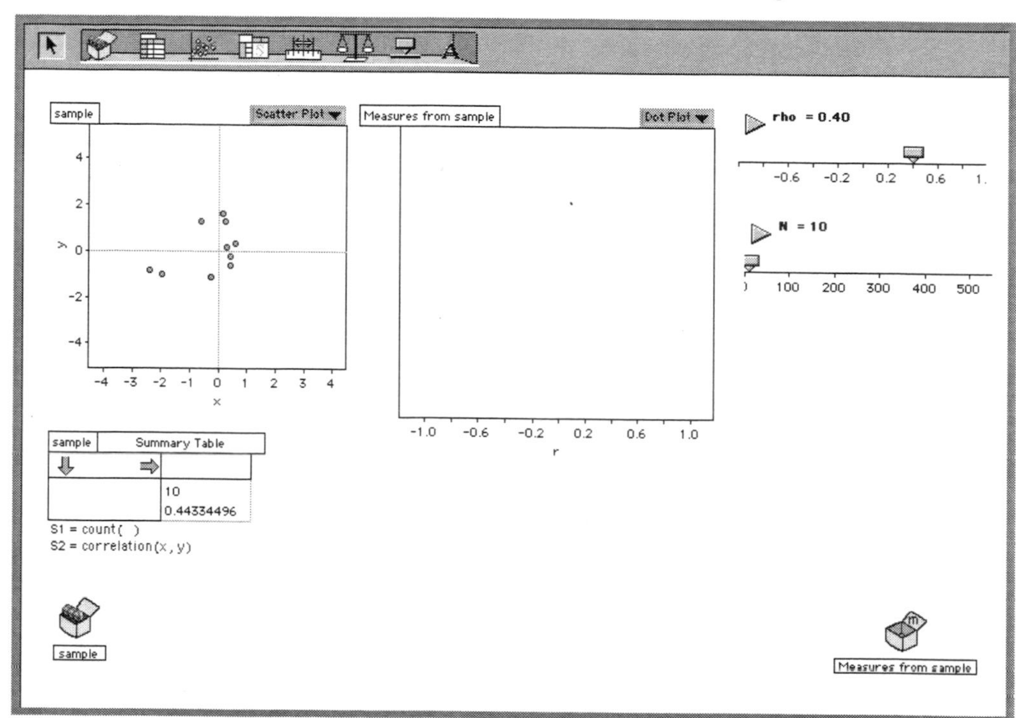

In the upper left, you see a graph of **N** points (here, **N** is 10, but in general, you control **N** with the slider at the right) drawn from a population of points that have a correlation coefficient of **rho** (here 0.40, see the slider at right again). Of course, this sample will probably not have the population correlation coefficient of 0.40. The *sample's* correlation (here, 0.44) appears in the summary table below the graph, as well as the number of points in the sample.

We'll get to the large, blank, central graph and its collection (**Measures from sample**, in the lower right) in a bit.

We begin by exploring.

The shortcut for **Rerandomize** is **clover-Y** on the Mac or **control-Y** in Windows.

- With nothing selected, choose **Rerandomize** from the **Analyze** menu to draw a new sample. See how the points change. Do so repeatedly; get a feel for the variation in the sample correlation. Do you ever see a *negative* correlation?
- Play with the **N** slider. Move it to about 200 points. Rerandomize some more to see how that correlation changes, and how the impression of correlation is more stable with a larger sample. Do you ever see a negative correlation?
- Play with the **rho** slider to see how different correlations look with 200 points.
- Then do the same with a smaller sample.

Now let's collect data on repeated samples. If we leave the settings constant and rerandomize, what will the distribution of correlation coefficients **r** look like?

You can type these numbers into the sliders to set them exactly.

- Set the sliders to **N = 30** and **rho = 0.4**.
- Click once on the **Measures from sample** collection (the box lower right) to select it.
- Choose **Collect More Measures** from the **Analyze** menu. Fathom will resample 40 times and display the correlation coefficients in the large, formerly empty middle graph.
- Repeat this procedure many times, for different combinations of **rho** and **N**. Be sure to use large and small samples; and correlations near zero, +1, and –1.

Questions

1. If you leave the correlation coefficient **rho** the same, what effect does **N** have on the center and spread of the distribution?
2. If you leave the sample size **N** the same, what effect does the correlation coefficient **rho** have on the center and spread of the distribution?
3. With a sample size of 10 and a true correlation of 0.5, about how well do you think you could know that true correlation based on a single sample?
4. How about with a sample size of 500?

A Really Cool Extension

Let's see a lot of these graphs side by side.

- Open the file **Correlation in Samples 2.ftm**. It will look something like this:

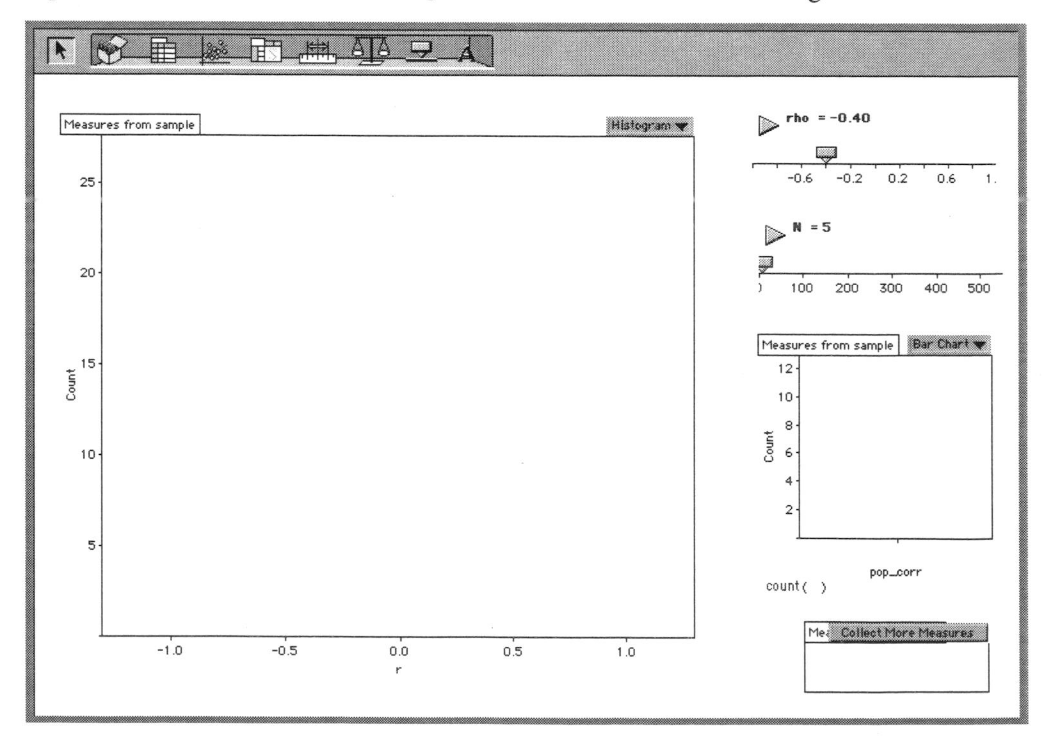

It looks practically empty, but really it's pretty much the same machinery as in the former file, just rearranged and slightly enhanced. You can see that we begin with **rho = –0.40** and **N = 5**. There is also a new bar graph (currently empty) of **pop_corr**.

- Press the **Collect More Measures** button on the collection in the lower right. The graph gets some data as before.

Try to predict what you will see.

- Change **N** to **25**. Do *so by typing in the slider, not dragging*; just edit the number and press **enter** when you're done.
- Press **Collect More Measures** again. Instead of replacing the data, Fathom adds the data to the graph. It will look like the illustration.
- Do the same for **N = 100** and **N = 500**. The graph now has four panes, one for each sample size.

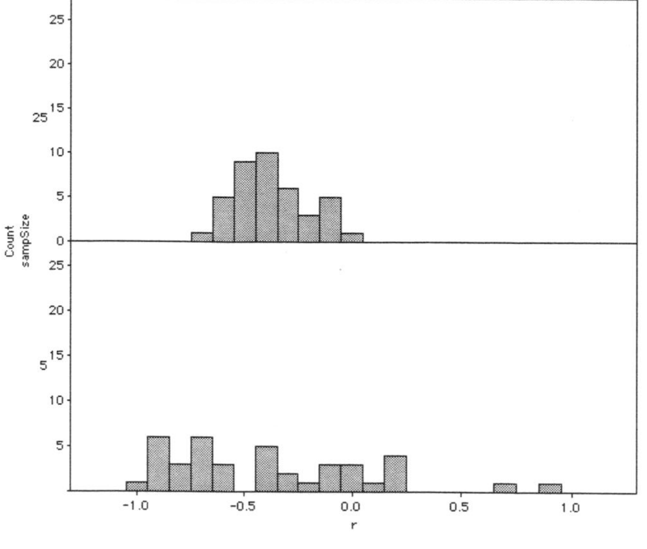

At this point, it should be clear from the graph that if you increase the sample size, you get a tighter correlation: you'll know the population correlation better based on a single sample.

If the histogram bars are a little chunky, adjust their widths, and maybe rescale the vertical axis. You can do this by dragging *carefully* or by double-clicking on an axis and typing into the ControlText you get. A bin width of 0.05 is fine.

- Now, leave **N** at 500 and we'll change **rho**. Set **rho** to 0.80. Press **Collect More Measures**. Fathom adds the points to the graph; now it has two spikes in the top pane.
- Work your way back down the sample sizes, leaving **rho** at +0.80, but collecting measures at **N = 100, N = 25**, and **N = 5**. Be sure to type in the numbers—don't just drag the sliders or you may be off by a bit.
- Now the *coup de grâce*. In the bar chart, click on the *bar* for **pop_corr = 0.8**. This selects all of the data that we took at that higher correlation, turning them all red in the other graph. You should see a graph like the one in the illustration.

See how the distributions in the **N = 5** case overlap? That means that if you see a sample correlation of, say +0.3, the population correlation could conceivably be anything from –0.4 to +0.8.

Another Question

web

5 It looks as if the spreads for **pop_corr = +0.8** are smaller than the ones for **pop_corr = –0.4**. Why would that be?

Demo 10: Regression Towards the Mean

Regression towards the mean • The meaning—and asymmetry—of the least-squares line

Let's use a context of test scores, as we did in "Standard Scores" on page 36. Suppose you do really well on the first test of the semester. Then the second test comes around, and you don't score as high. What happened? Were you just "due" for a bad score? Yes and no. You should rebel against that idea immediately because it sounds like the "gambler's fallacy"—the one where you expect tails after a long run of heads. However, it is true that in imperfectly-correlated data, if you're extreme on one measurement, you're likely to be less so in another. This is called *regression towards the mean.*

What To Do

- Open the file **Regression Towards the Mean.ftm**. It looks like this:

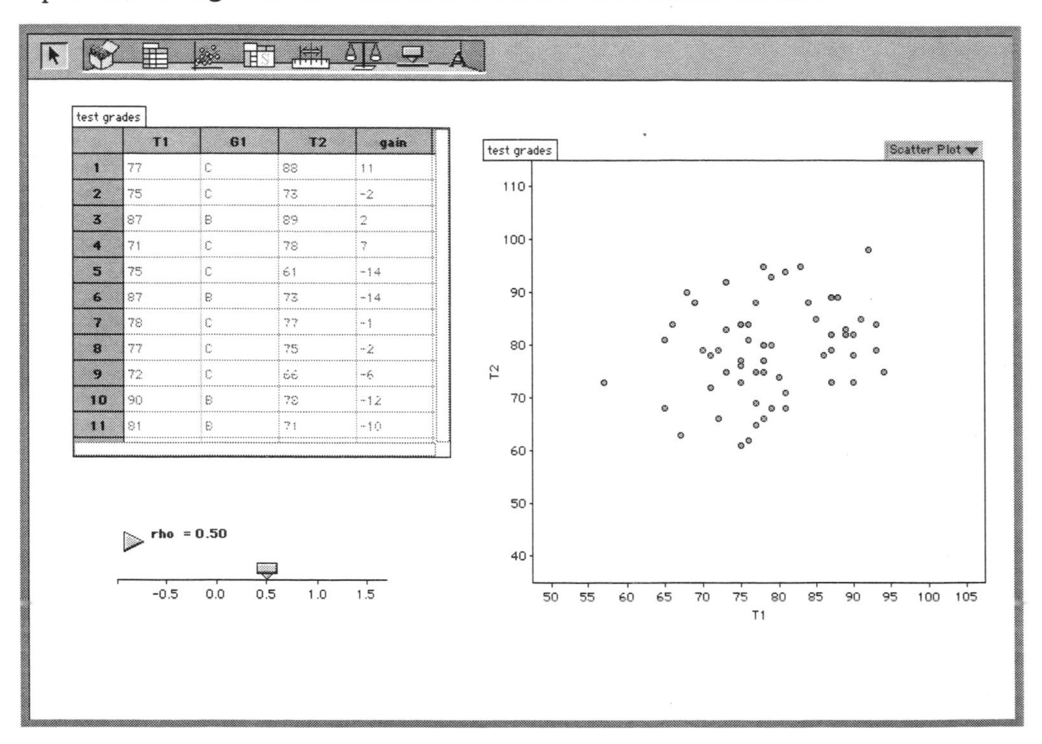

The case table shows the first 11 of 60 students from a fictitious class record book. **T1** and **T2** are scores on the first two tests; **G1** is the letter grade on that first test. The attribute **gain** is simply **T2 – T1**, the amount that student improved (or declined) in test score. The slider **rho** is the correlation between the two tests. The scatter plot shows the two test scores graphed against one another.

- Play with **rho** to see how it works; then reset it to about 0.5.
- Let's see what the grading scale is. Drag the attribute **G1** (grade on the first test) to the vertical axis of the graph, replacing **T2**.
- Let's also get some summary information on the graph. With the graph selected (it has a border), choose **Plot Value** from the **Graph** menu. The formula editor appears.

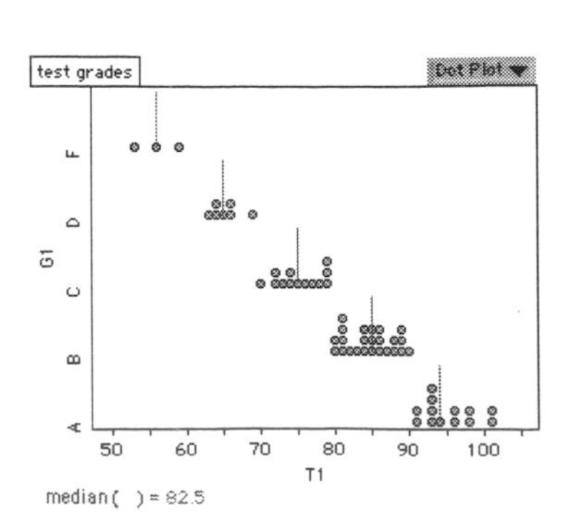

- Enter **median( )** and press **OK** to close the editor. Now lines appear showing the median score for each grade. Your graph should resemble the one in the illustration.

Note: throughout this demo, feel free to rerandomize the scores. It's quite possible, given random variation, that some of your graphs will not show what they're supposed to the first time. Just choose **Rerandomize** from the **Analyze** menu.

This graph shows the simple grading scheme: in the 90s for an A, etc.

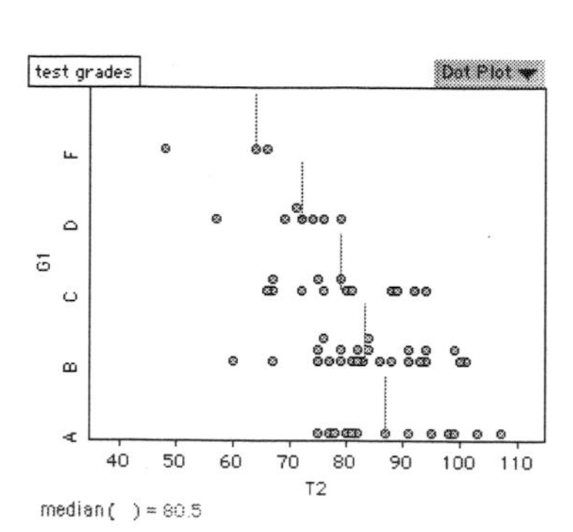

- Now drag **T2** to the *horizontal* axis, replacing **T1**. This graph shows you how the scores on the second test broke down when grouped by grades on the first test. You should be able to see that, as we would expect, students who did better on the first test generally did better on the second, though there are exceptions. Your graph should resemble the one in the upper illustration.

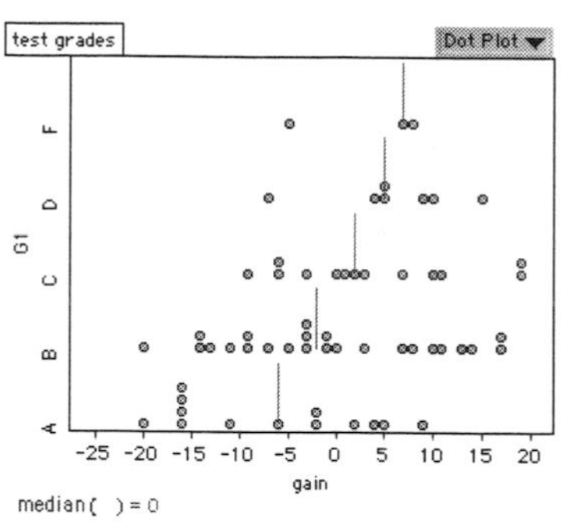

- But now drag **gain** to that horizontal axis, replacing **T2**. The trend reverses: now we see that those who got an "A" on the first test had the worst improvement. In fact, they have a median value of –6 points.

This is the phenomenon we're trying to demonstrate. Even though the higher-scoring students on the first test will generally score higher than others on the second, the performance of that *group* will generally decline. You see this phenomenon in many different settings. In one famous data set (I first saw it in *Statistics*, by Freedman, Pisani, and Purves), you can see that, while tall fathers generally beget tall sons, the tallest men tend to have sons shorter than themselves, while the shortest have taller sons.

Questions

1 What happens to this **gain** graph when you change **rho**?

web

2 When you flip coins and get six heads in a row, the chance of tails is still one half if it's a fair coin. Yet here, it looks as if when you get a good score you are more likely to get a lower score next time. What is it about this situation that makes it different?

Theory Corner

If you're wondering how we generated that correlated data, you can read about it in Fathom in "Simulating Correlated Data" on page 167, and why that works theoretically in "Correlated Data: Why the Way We Generate It Works" on page 177.

Extension

Let's look at this another way, connecting the idea of regression towards the mean with that of the least-squares linear regression line.

- Shrink the scatter plot and drag it to the left (covering the table if you wish) to reveal another, fancy scatter plot we have hidden behind the first one. Your document will now look something like this:

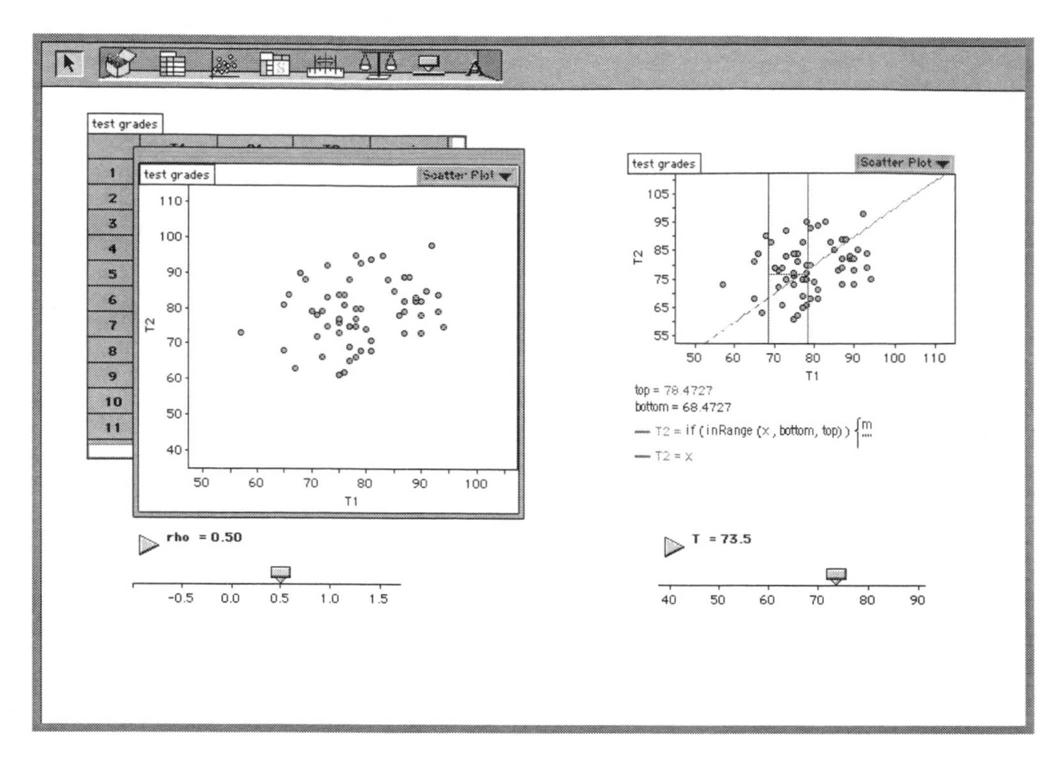

This new scatter plot—of **T2** versus **T1**, as we had at the beginning—now has four lines on it. The diagonal line is the line **T2 = T1**. That is, if you're below that line, your score got worse between the two tests. Two vertical lines delimit a region, and the short *horizontal* line between them shows the average score on **T2** for people between the lines. Finally, the slider **T** determines the center of the region; when you slide it, the vertical lines move.

- Move the slider. Watch the level of the horizontal line. In general, you should see that it is below the diagonal line for the high scores, and above it for the low scores—which is just what we saw in the other graph.
- One more thing: click on the graph to select it. Then choose **Least-Squares Line** from the **Graph** menu. The least-squares line appears.
- Now drag the slider again; in general, the horizontal line will match the least-squares line. That is, *the least-squares line is the line that goes through the mean value for each vertical strip of data.*

Another observation about this last graph: when the points form a characteristic "error ellipse" as they do here, the least squares line does *not* go along the ellipse's major axis. It's shallower. And if you switch the axes, the same effect occurs. How can that be? It's because the least-squares line is *asymmetrical*; here it's the regression of **T2** on **T1**, not the other way around.

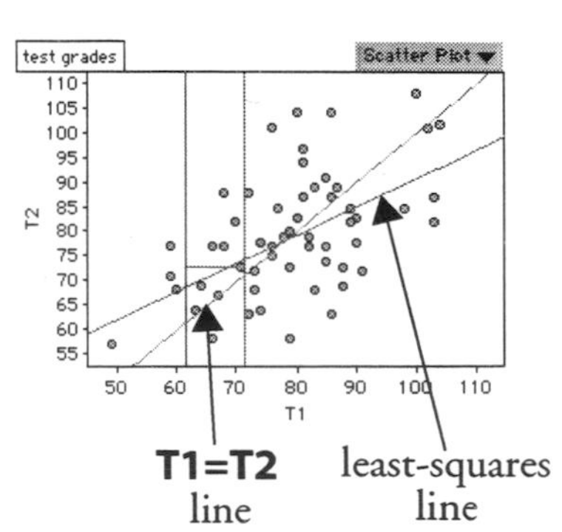

In contrast, the calculation of the correlation coefficient is symmetrical: the correlation of y and x is the same as that of x and y.

Random Walks and the Binomial Distribution

Many statistics books start their treatment of inference by looking at proportions. And when you do proportions, if you want to understand what's going on at the atomic level, you have to understand how binomial distributions work.

We aren't going to go all combinatorial here; you won't see a factorial in the place. Instead, we will look at some phenomena associated with the binomial distribution in their empirical rather than theoretical clothes.

In this section, you will find:

"Flipping Coins—the Law of Large Numbers" on page 48. With this demo, you can flip gazillions of coins and see how the proportion of heads approaches one half. Then you can do it over and over, and see the myriad paths that proportion takes, always to the same destination.

"How Random Walks Go as Root N" on page 51. If you take a random walk, on the average you'll wind up where you started. But the average distance you will be from the starting place increases as you keep walking; in fact, it increases without bound. Specifically, it is proportional to the square root of the number of steps. We'll see that happen in Fathom.

"Building the Binomial Distribution" on page 56. In this demo, you build the binomial distribution from the ground up—by sampling—and compare distributions with different population proportions.

"More Binomial" on page 59. Here we look at the binomial distribution slightly differently; in the previous demo, we looked at the distribution of proportions, but here we look at the distribution of counts as well. We also see how both change with sample size. You will see surprisingly subtle things; understanding them will help your understanding of the whole enterprise.

"Two-Dimensional Random Walks" on page 61. What if you walked randomly in two dimensions? We explore that here in Fathom. You'll see bizarre results for small numbers of steps that, like so much else in statistics, look smooth and normal with large numbers.

web

Remember: If you see web in the side margin, that means that there's a solution on the web at the time of publication. See **http://www.eeps.com**.

Demo 11: Flipping Coins—the Law of Large Numbers

How the proportion of heads approaches 0.5 as sample size increases • How the number of heads does not approach half the sample size

This demonstration shows how the proportion of heads you flip approaches the "true" probability of heads. First, we'll look at a file where you flip the coins one at a time...

What To Do

- Open the file named **Law of Large Numbers.ftm**. It will look like this:

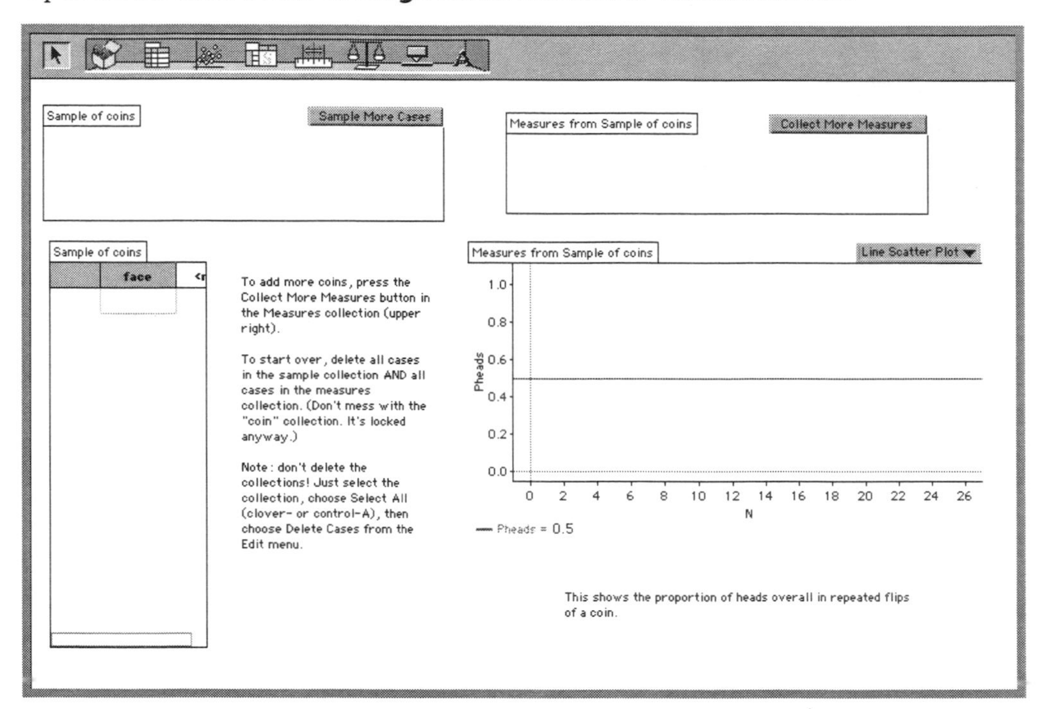

Here we have fairly complicated collections which start out empty. We will mostly watch the graph, which will show the proportion of heads for a number of consecutive flips. But we haven't flipped anything yet, so everything is empty.

To delete all of the cases in a collection, select an object of that collection (the collection, a case table, or even a graph); choose **Select All Cases** from the **Edit** menu; and then choose **Delete Cases** from the **Edit** menu.

- Click the button called **Collect More Measures** in the Measures (right-hand) collection. Fathom will flip a coin. You'll see the coin appear in the Sample collection and the cumulative proportion appear in the graph.
- Keep doing this and see how the proportion of heads tends towards 0.5.
- To start over, either re-open the file or delete all of the cases in the **Sample of coins** collection and all of the cases in the **Measures from Sample of coins** collection. *Do not delete the collections*—just the cases within them! (Note: for reasons that are really esoteric, Fathom's vaunted Undo feature does not work if you want to take back collecting measures. So that particular path—ordinarily one of the best—will not serve.)

We have built up the graph conceptually, starting with a single coin. Now, we will use fancy formulas to get faster performance. In our new file, we will use a collection that does its cases all at once instead of one at a time.

- Open the file named **Law of Large Numbers 2.ftm**. It will look something like this:

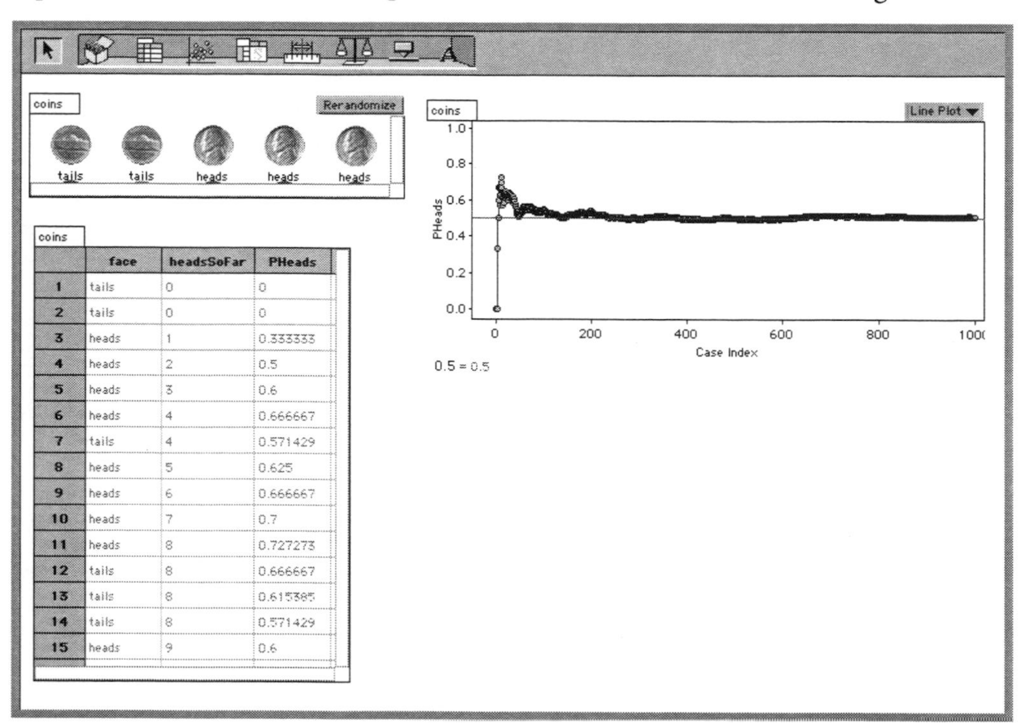

	face	headsSoFar	PHeads
1	tails	0	0
2	tails	0	0
3	heads	1	0.333333
4	heads	2	0.5
5	heads	3	0.6
6	heads	4	0.666667
7	tails	4	0.571429
8	heads	5	0.625
9	heads	6	0.666667
10	heads	7	0.7
11	heads	8	0.727273
12	tails	8	0.666667
13	tails	8	0.615385
14	tails	8	0.571429
15	heads	9	0.6

This shows how the proportion of heads (**PHeads**) changed over the course of 1,000 flips. In the upper left, you see the collection itself, and the first five coins. For example, here, you can see how **PHeads** is zero after one and two coins (they're both tails) and 0.333 after three coins (the third coin is heads). Below the collection, the case table shows you three attributes for the first 15 cases: the **face** of the coin, the number of **headsSoFar**, and **Pheads**, the proportion of heads in the collection so far.

- Click **Rerandomize** repeatedly to make Fathom flip the coins and redraw the graph.
- On the graph, zoom in to the first 50 or so cases and rerandomize a lot. A good way to do this, besides dragging the big numbers off the edge, is to hold down **option** (Mac) or **control** (Windows) and click near the zero in the horizontal axis. The axis will expand; with so many points, it may be a little slow.
- Zoom in to the *last* 100 or so cases—be sure to expand the vertical scale—and rerandomize a lot.

Questions

1 About what is the range of values you typically get for the proportion at 30 cases?

2 What is the range after 1000 cases?

web

3 Why is graph so much flatter at the end than it is at the beginning?

Extension

- Restore the scale of the graph by re-choosing **Line Plot** from the menu in the corner of the graph.
- Make a new graph by dragging one off the shelf. Make it about the same size and shape as the existing graph, if you can.
- Double-click the collection (the box with the coins in it) to open its inspector. Note: do your double-clicking down where the coins are—you can even double-click a coin—

rather than up between the name and the **Rerandomize** button. That zone "between the ears" is not really part of the collection.

- Click on the **Cases** tab to make the **Cases** panel appear.
- Drag the last attribute, **headsExcess**, to the *vertical* axis of the new graph. It will make a dot plot.
- Close the inspector.
- Change the graph to a line plot by choosing **Line Plot** from the popup menu in the graph itself. Your screen will look something like this:

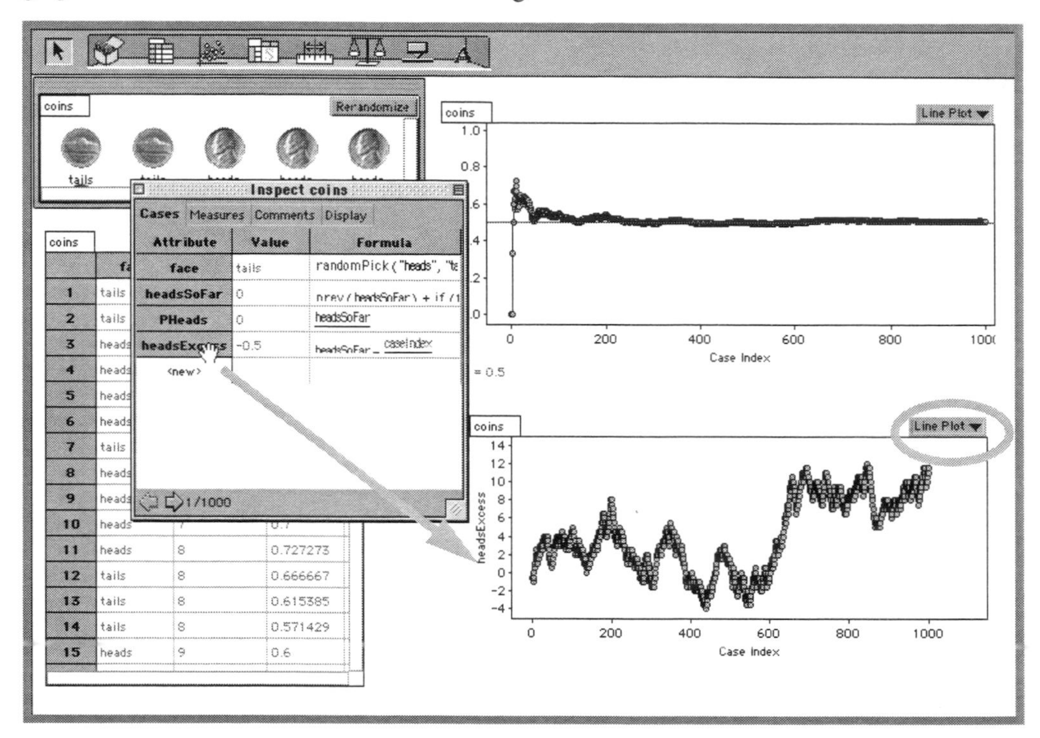

This new attribute, **headsExcess**, is how many extra heads you have flipped so far, as compared with exactly one-half of the coins. So, for example, if the first coin is tails, you have –0.5 excess heads: you're "expecting" one-half of a head, and you have none. So your "excess" is –0.5. You can expand the case table to see this attribute as well—look at it until you understand what it does.

- As before, repeatedly rerandomize, zooming in to various parts of the graph. You'll discover that you need vertical bounds running from about –40 to +40 to get all the points in reliably. Compare this graph with the one for **PHeads**.

More Questions

1. What range of values do you get for **headsExcess** after 1000 cases?
2. What is the range after 30 cases?
3. Why is this graph no flatter at the end than it is at the beginning (in contrast to the **PHeads** graph)?

Demo 12: How Random Walks Go as Root N

How the distance from the origin increases with the number of steps

The previous demonstration showed how the proportion of heads you flip approaches the "true" probability of heads. This one shows how the number of heads gets farther and farther from the expected value. Learning why these two apparently contradictory statements are not in fact contradictory is one of the most important things you can do to understand statistics.

Instead of staying with flipping coins, we'll use a slightly different context—a random walk. The idea is that, for each step, you randomly choose whether to go east (to the right, positive) or west (to the left, negative). After *n* steps, how far have you gone?

On the average, you will not go anywhere. Your mean position will be where you started. But it is also true that after many steps, you're very unlikely to wind up *exactly* where you began. So how far—in absolute terms—are you likely to be from the origin after *n* steps? We will simulate this situation in this demo to find out. And the answer is that *the distance is proportional to the square root of n.*

Note: In this demo, we will turn the graphs on their sides, so that the "time" axis will go up instead of to the right. This will fit with our left/right walk metaphor; it will also map well onto the graphs of the binomial distribution we will create in "Building the Binomial Distribution" on page 56.

What To Do

- Open the file **Random Walk Root n.ftm**. It will look something like this:

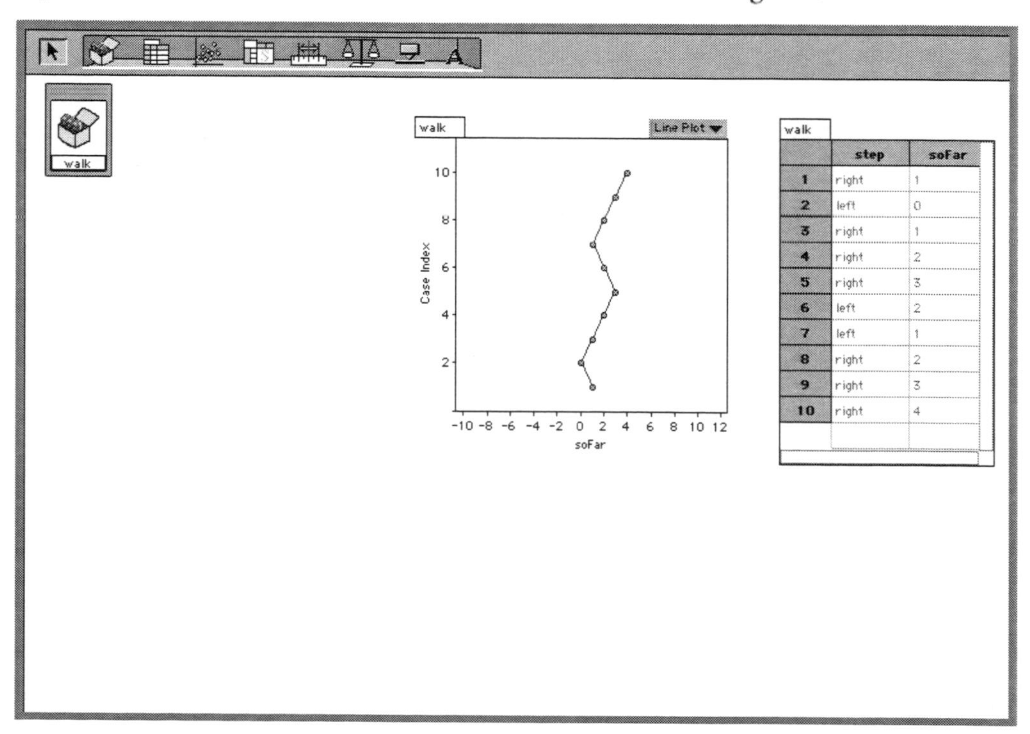

	step	soFar
1	right	1
2	left	0
3	right	1
4	right	2
5	right	3
6	left	2
7	left	1
8	right	2
9	right	3
10	right	4

Here, the collection **walk** contains one random walk, shown in the graph. It starts at the bottom; it happens that after 10 steps, we've landed on +4. First, we'll see how the walks change…

- Double-click the **walk** collection to open its inspector. Then choose **Rerandomize** from the **Analyze** menu.

The shortcut for **Rerandomize** is **clover-Y** on the Mac or **control-Y** in Windows.

- Rerandomize repeatedly to see how the walk changes. You can also see, in the **Measures** panel of the inspector (shown), how several quantities change: the **end** position of the walk; the square of that (**endSq**), and its absolute value (**endAbs**).

Inspect walk

Cases | Measures | Comments | Display

Measure	Value	Formula
end	4	count (step = "right") – count (step = "left")
N	10	count( )
endSq	16	end^2
endAbs	4	|end|
<new>		

Note: You can also see their formulas.

Wouldn't it be nice to record all those numbers somewhere? That's a job for a measures collection—which has really been there all along, hidden.

- Close the inspector to make more screen space.
- Choose **Show Hidden Objects** from the **Display** menu. Now the screen should look something like this:

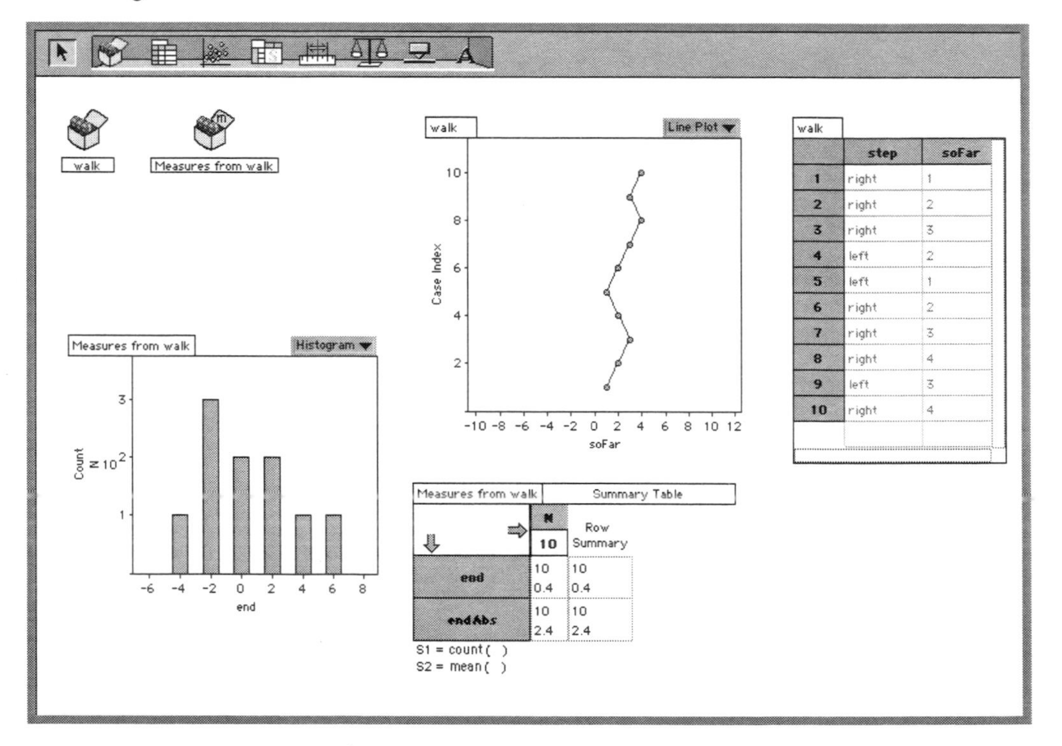

Three new objects appeared: a measures collection (**Measures from walk**), a graph, and a summary table. The measures collection already holds the results from 10 walks.

See in the graph that the ending positions (**end**) among the 10 walks ranged from –4 to +6. See in the summary table that the average position at the end is 0.4, but the average *absolute* position at the end was 2.4.

- Click the measures collection (it looks like the picture at left) once to select it. Choose **Collect More Measures** from the **Analyze** menu. Fathom will produce a new walk and add its results to the measures collection. The graph and the summary table will update to show the results from 11 walks instead of 10.
- Keep collecting more measures this way until you have at least 20 walks. Clearly we need a faster way to get these walks done.

- Double-click the measures collection to open its inspector. It will open to the **Collect Measures** panel, as shown. Edit the number so that you will collect 80 measures instead of one. (It doesn't have to be exactly 80. Just a lot.)

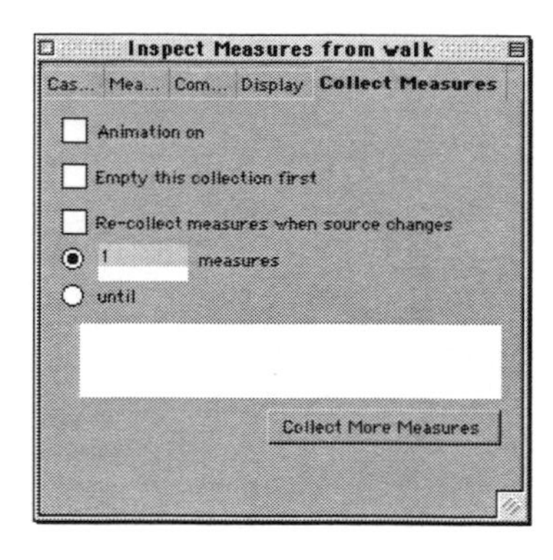

Note: depending on your screen, it may not be obvious where to put this. If you're lucky, it will appear directly over the graph of the single walk—top center. If it's not, move it there.We don't need that graph now.

Having exactly 500 is not necessary, but it's nice because then the decimal values in the summary table will be *short*—you won't have to scroll to see the data.

- Click the **Collect More Measures** button. Fathom will update everything. Keep collecting more measures until you have at least 500 measures from walks of 10 steps each.
- Now we need a different number of steps. Click once on the **walk** collection (the one in the upper left) to select it.
- Choose **New Cases...** from the **Data** menu. Give the collection 10 more cases (for a total of 20 steps).
- Set up the measures inspector to collect 500 measures, then click the **Collect More Measures** button. Again, Fathom updates, but this time you'll see data for 20 steps (**N = 20**) separate from the data for 10 steps. The summary table will look something like the one at right.

Measures from walk — Summary Table

	N = 10	N = 20	Row Summary
end	500 -0.136	500 0.012	1000 -0.062
endAbs	500 2.488	500 3.38	1000 2.934

S1 = count( )
S2 = mean( )

- Keep adding steps and collecting measures[1] so you have data for 10, 20, 40 80, and 160 steps. Stretch the summary table as it grows, making it wider so you can see the data. You may need to scroll.
- If you like, drag **endAbs** to the horizontal axis of the graph, replacing **end** to see what its distributions look like.

Questions

1. How does the average position at the end of the walk (called **end** in the summary table) change as you increase the number of steps?

web

2. How does the average absolute value at the end of the walk (**endAbs**) change?
3. Why are these two quantities different?
4. How does the shape of the graph change as you increase the number of steps?
5. Judging from the summary table, is the mean absolute distance proportional to the number of steps? How do you know?

Onward!

Now we want to graph the numbers in the summary table. We have set all this up for you and generally cleaned up the screen. We begin with a new file, based on the previous one:

1. Note: this is a little tricky. You're alternating between adding cases to the **walk** collection (to make a walk of more steps) and adding cases to the **Measures from walk** collection (to get the results of 500 walks). It's easy to mess up, but also easy to recover if you keep your eyes open. If you get too many cases in the **walk** collection, i.e., too many steps, select the excess in the graph or the case table and choose **Delete Cases** from the **Edit** menu. Don't worry about getting too many measures, but if you get some cases for an **N** you don't want, select them in the lower-left graph and, again, choose **Delete Cases** from the **Edit** menu.

- Open the file **Random Walk Root n part 2.ftm**. It should look like this:

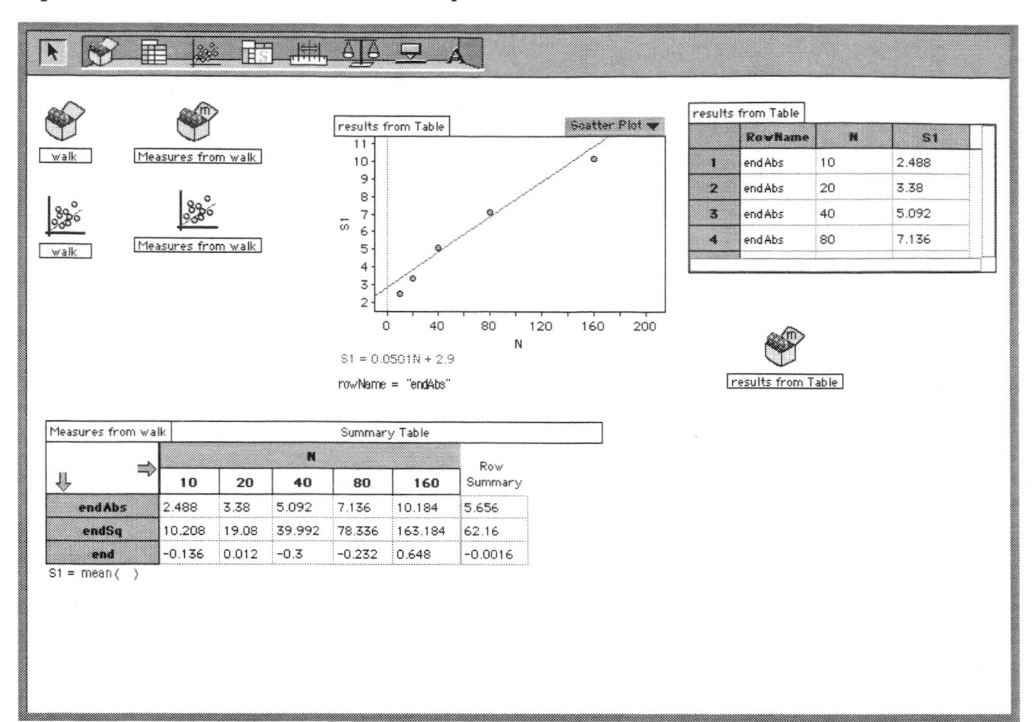

The old graphs are collapsed into icons, and the summary table, at the bottom, is streamlined (we have removed the **count( )** formula—there are 500 cases in each column). And we have added the **results from Table** collection, which is really a measures collection from the table itself. The graph in the middle shows how the mean absolute distance changes as a function of the number of steps.

Note: We control what to plot with the *filter* (**rowName = "endAbs"**) at the bottom of the graph. The actual attribute is that thing **S1** from the table at the bottom, which is the mean. But the filter tells the graph that you only want to see the absolute values. One can edit the filter by double-clicking it, for example, to **rowname = "endSq"**.

This whole setup is complex and layered. Be sure to understand that the three collections are linked left to right:

- **walk** is a collection of steps—at this time, 160 of them, which is where we left off. Each case is one step, either left or right. To see the last walk, open the graph below it by dragging the lower-right corner. (Then shrink it again.)
- **Measures from walk** is a collection of data from 2,500 walks—500 at each of five different lengths. Each case is a walk, and its attributes are the number of steps, the position at the end, and that position's absolute value and square. Open the graph below it (drag the lower-right corner) to see the distribution of final positions.
- **results from Table** is a summary of the 2500 walks. It shows the mean of the absolute value (and square, and original **end** number) of the number of steps in the 500 walks at each of the five lengths.

To add cases to the **walk** collection, select it by clicking on it once, then choose **New Cases...** from the **Data** menu.

- Optional: add another 40 cases to the **walk** collection (for a total of 200) and **Collect More Measures**. The table and graph will update. You'll need to stretch the table.
- Drag parts of the brown movable line in the graph to see if you can make it fit the data points there. It should be clear looking at the line that the pattern of the points is genuinely curved, which means that the mean absolute value of the distance is *not* proportional to the number of steps.
- Choose **Show Hidden Objects** from the **Display** menu. A new graph appears, showing the mean *square* distance as a function of the number of steps.
- Drag the line to fit the data. It does a lot better.

We have shown that, in a series of random walks, the number of steps you take is proportional to the mean of the square of the distance from where you started. In fact, the slope of the line is 1 and the intercept is zero, that is,

mean square distance = N

Harder Stuff

1. Edit the filter on one of the graphs to show how the mean of **end** does not change much as you change **N**.
2. Edit the formula in the table (or add a formula) to see how the *medians* (instead of the means) of these quantities change. Would you rather use median here? Why or why not?
3. Mean absolute deviation (which is what we were looking at, the mean of **endAbs**) ought to be a measure of spread. So is standard deviation. Figure out how to compare standard deviation to the mean absolute deviation. Are they proportional? If so, what is the constant of proportionality? Explain these results.
4. In what respects is this random-walk situation is identical to (fancier word: *isomorphic* to) the coin flipping we did in "Flipping Coins—the Law of Large Numbers" on page 48?
5. Explain clearly why, if these two demos are isomorphic, the spread decreased with n when we flipped coins, but increased with n when we did a random walk.
6. web — If the mean square distance is N, does that mean that the mean *absolute* distance is the square root of N? Exactly? Not at all? Only in the limit? Only on Tuesdays? (One way to look at it: in the case table, create a new attribute equal to the square root of **N**; then replace **N** with this new attribute in the first (**N-endAbs**) graph. The line is straight; what is its slope? How do you explain that it is not 1.0?)
7. Suppose you took a long random walk and plotted the number of steps against the square of the distance. Will the graph approach a straight line? Explain.

Theory Corner

See "A Random Walk: Two Proofs That the Mean Square Distance is N" on page 176.

Demo 13: Building the Binomial Distribution

Constructing the binomial distribution by resampling • How the distribution depends on the population proportion

The binomial distribution is at the root of statistical situations having to do with proportions. It can be confusing to remember just what it's a distribution *of.* The conceptual problem is in the layering of the situation: you pull a card from a deck of red and black cards—where's the distribution? The answer is that you have to draw *n* cards (with replacement) and count the black ones; and *then you have to repeat that process.* The distribution is the distribution of *numbers* (or, equivalently, proportions) of blacks in those repeated series of draws.

This demo tries to clarify this "layering" and to give you some mental images of the distribution.

What To Do

- Open the file **Building Binomial.ftm**. It will look like this:

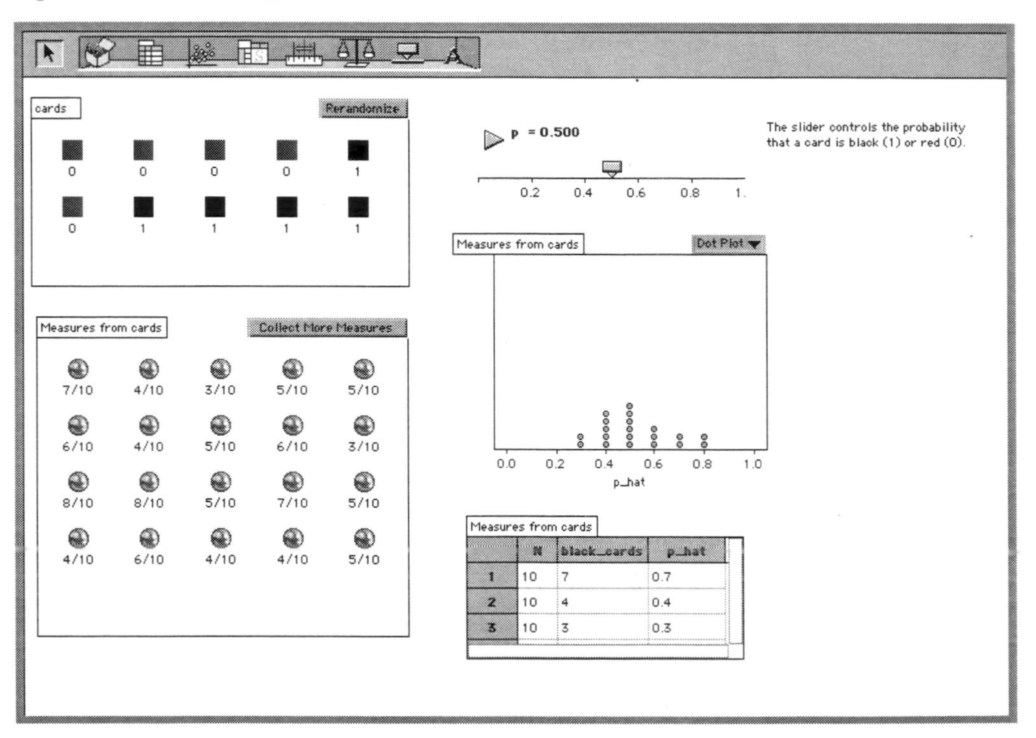

A collection, **cards**, shows ten red and black cards, coded 0 and 1. The probability that a card is black is controlled by the slider **p** top center. Below it we see the results of 20 sets of 10 cards. You can see a graph of those results (shown as proportions) and the top of a table as well. Note that we are not simulating the *whole* deck here; the **cards** collection is already a sample, drawn randomly from the deck.

We begin by focusing only on the cards and the slider.

- Click the **Rerandomize** button in **cards** repeatedly. See what happens.
- Drag the slider; see what happens (nothing will change down below).

You should see that the slider controls the probability that a card is black, but that the number of black cards varies from set to set. That is, even though the probability might be 0.500, there aren't always exactly 5 cards.

So now we want to collect these numbers of black cards to see how they vary.

- Reset the **p** slider to 0.5.

- Click the **Collect More Measures** button on the lower (**Measures from cards**) collection to see what happens. See that, when the collecting is done, the last gold ball corresponds to the current **cards** collection.

Fathom empties the measures collection and then collects 20 new measures, rerandomizing the data collection each time. The "flying balls" show how each new measure (which appears as a gold ball with a fraction under it) represents an entire new set of cards. Note that when it starts over, Fathom rescales the graph.

Note: we have animation turned on in order to show this process more slowly. You can, of course, turn it off when you're ready. But if the process is too fast, here's a suggestion: change the measures collection to collect only one measure at a time, without emptying the collection. Then you can look at each **cards** collection and see what new measure appears.

- To make it easier to compare graphs from one setting to another, rescale this graph manually so that it goes roughly from 0 to 1. (See"Rescaling Graph Axes" on page 18.)
- Click on a gold ball in the collection to see where it is in the graph. Then select points in the graph to see which balls they correspond to (e.g., drag a rectangle around all the 0.5's).
- Drag the attribute **black_cards** from the case table (at the bottom) to the horizontal axis of the graph, replacing **p_hat**. See how the two graphs show the same distribution. Put **p_hat** back when you're done.
- Explore what happens when you change **p** and then press **Collect More Measures**.

Questions

1 As you change **p**, how does the distribution change?

2 As you re-collect measures (leaving **p** constant), how does the distribution change?

3 If you set **p** to 1.0, what will the distribution look like? Why?

The distribution of **black_cards** is called a *binomial* distribution. The "bi" part comes from "two." There are two possible outcomes for each card: red and black. It could be heads or tails, success or failure, male or female, east or west, whatever. The probability of each choice can range from 0 to 1; if you know one probability (p), the probability of the other (often q) is one minus the probability of the first (i.e., $q = 1 - p$).

The distribution shows the proportions you get for one of these choices when you draw N cards repeatedly. These proportions (often called p-hat, or $\hat{p}$—the sample proportion) tend to be close to the true probability, but there's variation about that "expected" number.

If you do a random walk repeatedly, the distribution of ending positions is also binomial. And if you have a large enough sample size, the binomial distribution looks very much like a normal distribution (given certain conditions; see "Why np>10 is a Good Rule of Thumb" on page 107).

Onward!

Now we'll study more explicitly how the distribution depends on the value of **p**.

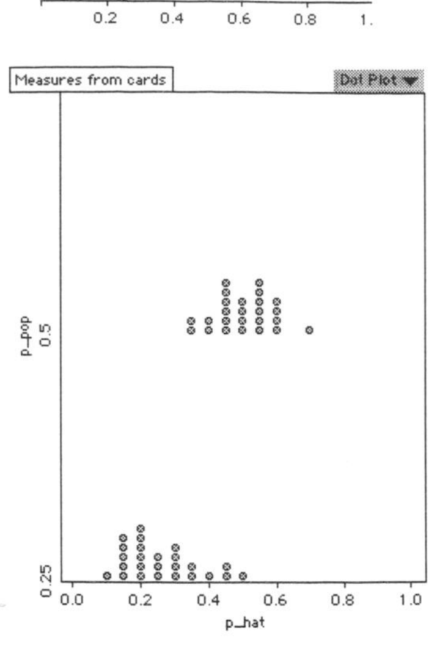

- Open the file **Building Binomial part 2.ftm**. It will resemble the previous file, but there are now 20 cases in the **cards** collection instead of 10, and 25 measures (gold balls) instead of 20. We have also made the graph bigger and made changes to the way the measures collection collects. (So *don't* click **Collect More Measures** until we tell you to!)
- Play with the slider to see what it does to the **cards**. When you're ready, set it at 0.25 (by editing the number in the slider).
- Now click the **Collect More Measures** button. Fathom quickly collects more measures, *adding* them to the collection—instead of replacing them—and adding the points to the plot.

Your graph should look something like the one in the illustration, split to show how the distributions are different. We have scaled the axis to show the range 0 to 1. The attribute **p_pop** (probability for the population) is the value of **p**, the slider, at the time the measures were collected.

- Set the slider to three more values and collect measures each time so you have a total of five different probabilities. An interesting set of values for the slider is {0.1, 0.25, 0.5, 0.8, 0.95}.
- Change the graph to a histogram, an Ntigram, and a box plot using the pop-up menu in the graph to see other representations of this distribution.

Questions

1. How does the center of the distribution depend on **p_pop**?
2. How does the spread of the distribution (for example, the width of boxes in the box plot) depend on **p_pop**?
3. Why should the spread be smaller when **p_pop** is close to 1 or 0? web
4. Why is it that before, all the proportions in the graph were 0.40, 0.50, 0.60, etc., while now we have 0.45, 0.55, and the like?

Extensions

1. Select the graph and choose **Plot Value** from the **Graph** menu. Use the formula editor to specify a value to plot; Fathom will plot that value for each part of the graph. The simplest is **mean( )**, but you could, for example, plot the 10th percentile with **percentile(10,)**. (Leaving the blank after the comma tells Fathom to use the attribute on the axis.) You can also plot multiple values by choosing **Plot Value** again. Edit an existing formula by double-clicking the formula at the bottom of the graph.
2. We have just explored how the distribution depends on **p_pop**. Use **Building Binomial part 3.ftm** to see how it depends on the sample size, **N**. To change the sample size, select the **cards** collection and then choose **New Cases...** from the **Data** menu. (It starts out with 10.) An interesting set of values for *N* is {10, 40, 100, 200, 1000}. This is closely related to "How Random Walks Go as Root N" on page 51.

Demo 14: More Binomial

How the binomial distribution depends on sample size for small N • The relationship between the distribution of sample proportions and the distribution of sample counts

In this demo, we're not going to make the samples and then count up how many successes there are in each set of draws. Instead, we'll use Fathom's built-in **randomBinomial( )** function to skip all that (therefore, if you have any doubts about what's going on, you should look at "Building the Binomial Distribution" on page 56).

In this way we'll take a different look at what's confusing: how is it that we can take this yes/no kind of event—a binary event—and turn it into a distribution that's pretty smooth and has a hump in the middle? This time, we'll use Fathom's dynamic sliders to control the *sample size.*

What To Do

Open the file **Binomial Small N.ftm**. It will look something like this:

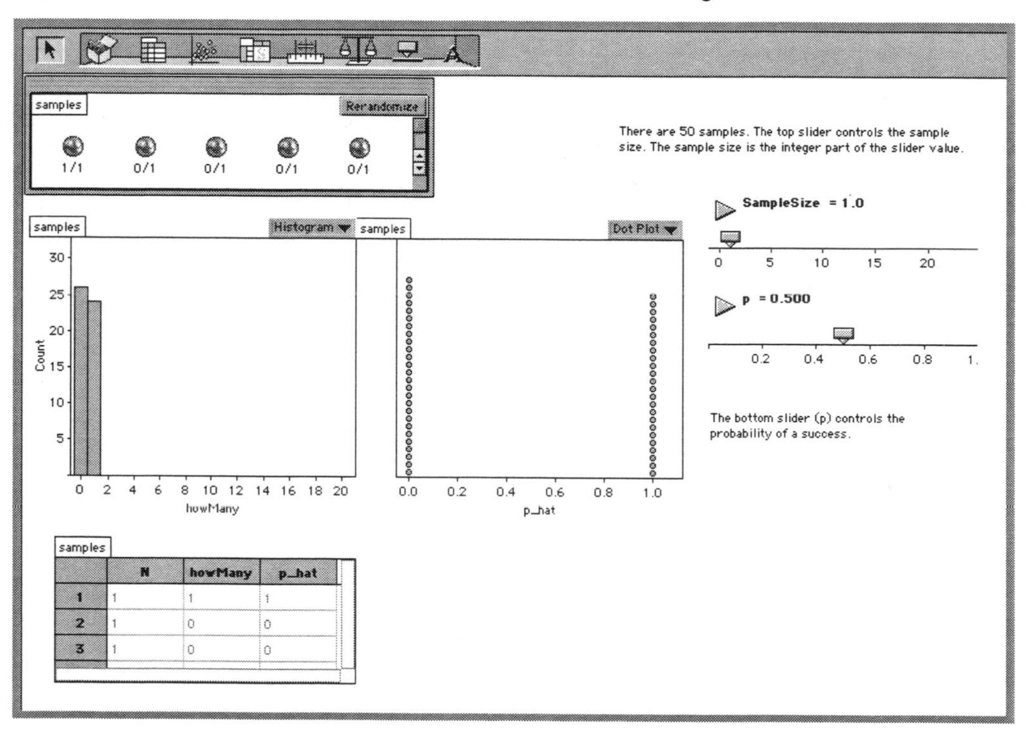

Each case in the collection represents one sample; the caption under each case—each gold ball—tells how many "successes" there were in how many trials. This could be heads in flips of the coin, black cards in a set of draws, whatever. The graph below left shows a histogram of the counts. At right, you can see two sliders with explanatory text. The integer part of **SampleSize** is the sample size[1]. Below the graph you can see the first three cases in the data collection. There are 50 cases altogether, that is, 50 samples of size one.

In the picture, our one object has a 50% chance of success. So in our 50 cycles, we see about 25 cases with zero successes and about 25 with one success, which is what you'd expect. In the graph to the right of the histogram you see the same data represented as proportions instead of absolute numbers. The possible values with one card are **p_hat = 0** and **p_hat = 1**.

1. We would have used **N**, but we're using that for the attribute in the case table. So the *slider* is **SampleSize**, the *attribute* is **N**. They mean the same thing conceptually.

- Click the **Rerandomize** button on the **samples** collection in the upper left. (Or choose **Rerandomize** from the **Analyze** menu.) Observe what happens.
- Move the slider named **p** that controls the probability of a success. See what that does to the two graphs. Reset that probability to about 0.5.
- Finally, gently move the **SampleSize** slider. Every time it passes an integer, it increases the sample size by one. Watch what happens to the two graphs. Try especially to understand what happens when the sample size is 1, 2, 3, and 4—and then how it gets to look as it does up around $n = 20$.

It would be great if Fathom had an "integer slider"—and it will eventually. For now, the attribute **N** is the integer part of **SampleSize**—and that's what really drives the simulation.

- Try changing sample size for different values of **p**. See what happens.

It's easy to look at the **p_hat** graph for $n = 1$ and the one for $n = 20$ and think that they are two completely different animals. One is grossly bimodal, just two spikes at opposite ends. The other is a classic distribution, practically normal, with *nothing* at the ends. If you stare at that **p_hat** graph as you move the **SampleSize** slider, it can be a little hard to make sense of.

On the other hand, looking at the **howMany** graph—the one in absolute number instead of in proportion—may help clear it up. After all, when $n = 1$, it's just a big spike, over on the left. As you add sample size, the hump gets less steep, and moves to the right. The process is continuous, not a strange transformation from one kind of graph to something completely different. This is, I suppose, an advertisement for looking at a variety of representations in order to understand something.

Questions

1. In that first step, when you clicked **Rerandomize**, what happened? How much variation was there in the number of "0" and "1" cases?
2. Consider the case where **SampleSize** = 1 (as at the beginning) but where **p** has been changed. How could you estimate **p** using only the graphs?
3. What happens to the graphs when you add one more to the sample size?
4. How can you predict where the peak will be in the **howMany** graph? How about the **p_hat** graph? (web)
5. How do the two graphs relate to one another?
6. What is the algebraic relationship among **howMany**, **N**, and **p_hat**?

Extension

You can put the theoretical binomial function on the **howMany** graph like this:

- Click on the graph to select it.
- Choose **Plot Function** from the **Graph** menu.
- Enter **binomialProbability(round(x), last(N), p)*count( )**. Press **OK** to close the formula editor. The function appears; the data should roughly match it.

Why those weird arguments? That function, **binomialProbability**, takes 3 (or more) arguments: the number you're interested in *x*, the sample size, and the probability. So **binomialProbability(3, 4, .5)** is the chance that you get exactly 3 successes in 4 tries with probability 0.5. But the first two have to be whole numbers, otherwise it doesn't tell you anything. So we have to round and do other shenanigans to avoid decimals.

Demo 15: Two-Dimensional Random Walks

Unexpected behavior in 2D random walks • How the 2D walk eventually looks like a 2D normal distribution

This demo may be beyond the scope of most introductory statistics curricula; it is in here partly just for fun because we'll see something that is unexpected (at least by this author). But it also helps us think about one-dimensional random walks.

There are several ways you could imagine a two-dimensional random walk. In one way—let's call it Cartesian—you flip a coin for the *x*-axis and a coin for the *y*-axis. Then you take one step along each axis, randomly positive or negative depending on the coin.

Instead, we'll do a *polar* random walk. We'll choose a direction at random, and then take one step in that direction.

What To Do

- Open the file **2D Random Walk.ftm**. It will look like this:

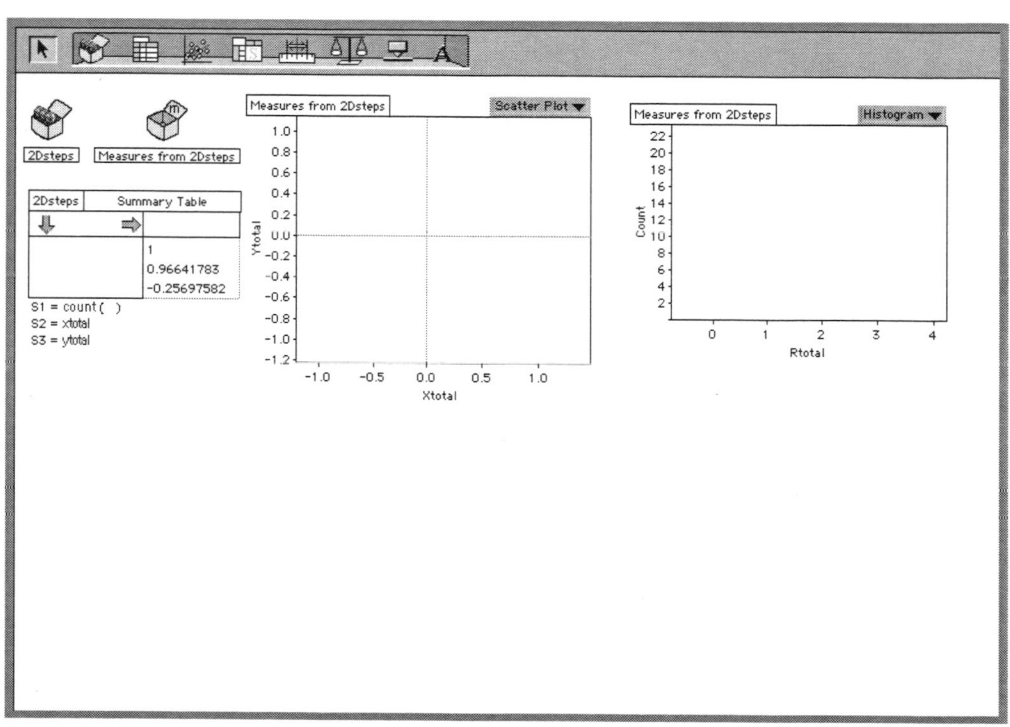

In the upper left, you see the collection (box with gold balls) called **2Dsteps**. This contains the individual steps of our walk. The summary table below it shows the number of steps (one at first) and the ending points (**xTotal** and **yTotal**) of the entire walk.

The collection and graphs to the right will collect and display the results from many walks. Let's make those appear.

- Click once on the **Measures from 2Dsteps** collection to select it.
- Choose **Collect More Measures** from the analyze menu. A point appears in the scatter plot—it's the endpoint of the walk.
- Continue to collect more measures in this way, one point at a time, until you see the shape of the graph. It should be a circle—and that should make sense.

The shortcut for **Collect More Measures** is **clover-Y** on the Mac; **control-Y** in Windows.

Note: The far right graph will probably look a little strange; it should be a spike at **Rtotal** = 1. It's a graph of the distances of the ends of the walks from the origin. In this case, all of the values should be equal to one; in practice, they won't be *exactly* equal to one due to roundoff error. You can re-choose **Histogram** from the pop-up menu in the graph to make it try to rescale itself. Just remember to select the **Measures from 2Dsteps** collection again if you want to collect more measures.

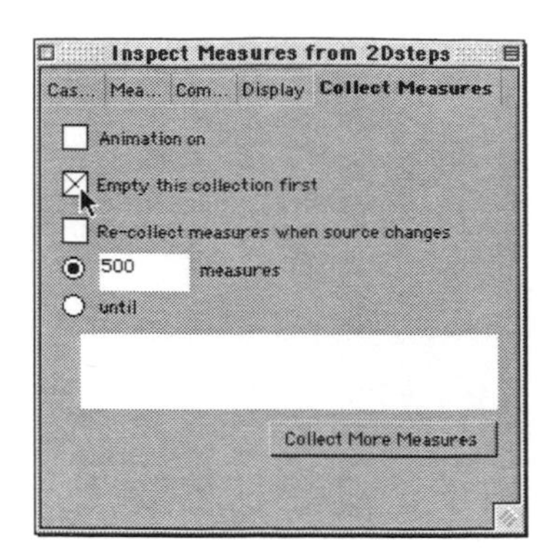

- Let's speed things up. Double-click that measures collection to open its inspector, and press the **Collect Measures** tab to bring up that panel.
- Change the settings in the panel to collect 500 measures at a time, turn off the animation, and set it to **Empty this collection first**, as shown.
- Now, press the **Collect More Measures** button (it's the same as selecting and choosing the menu item, but there's enough screen space to leave the inspector open). After a bit, the graph will update to show you a nearly complete circle.

Now the real fun begins. We are going to make our walk longer: instead of our boring one-step random walk, we'll take a *two*-step random walk! Whee-ha!

- Click the original, **2Dsteps** collection to select it.
- Choose **New Cases...** from the **Data** menu. Add one case and press **OK**. The table should update to indicate that you now have two steps in your walk.

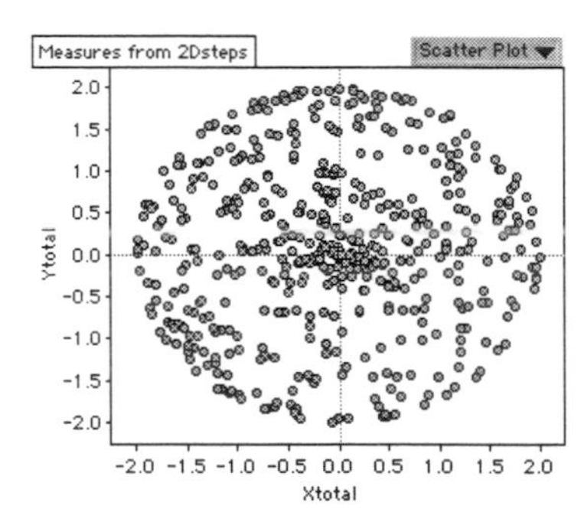

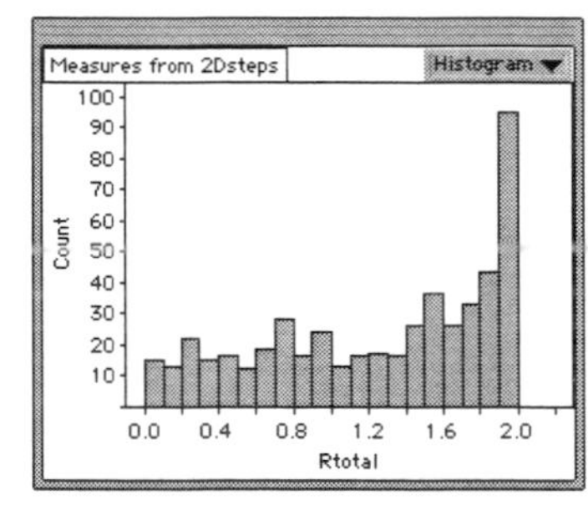

- Press the **Collect More Measures** button (or select the measures collection and choose the item from the **Analyze** menu). The graphs update to show you the results of 500 two-step random walks. (You may need to re-choose **Histogram** from the popup menu in the right-hand graph to get it to scale properly.)

Notice two things about the distribution of endpoints for these walks: first, there is a tight cluster near the origin; second, the histogram shows a sharp upswing as you get to a radius of 2. Think about what those walks must be like.

Note also: if there is any confusion about the correspondence between the two graphs, click on one of the bars of the histogram to select it.

Harder Stuff

1. It should be pretty clear that, since each step is only one unit long, all points must lie in a disk defined by $r \leq 2$, where r is the distance to the origin. What would the two graphs look like if the distribution of points were *uniform* in that disk?

2 Use that result (or anything else you can think of) to explain why the obvious clump near the origin doesn't show up in the radial histogram, and why the "density ring" at $r = 2$ is so striking in the histogram but not in the scatter plot.

Onward!

- Using the same procedure we used above, add a third case to the **2Dsteps** collection. That is, we will now have a three-step walk.
- **Collect More Measures** again. You should see graphs like the ones in the illustration.

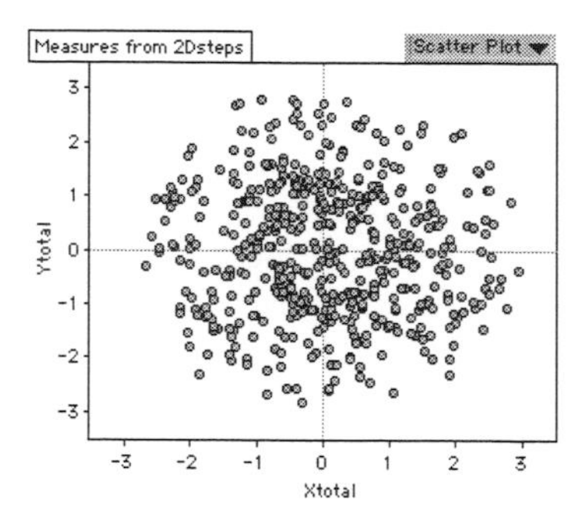

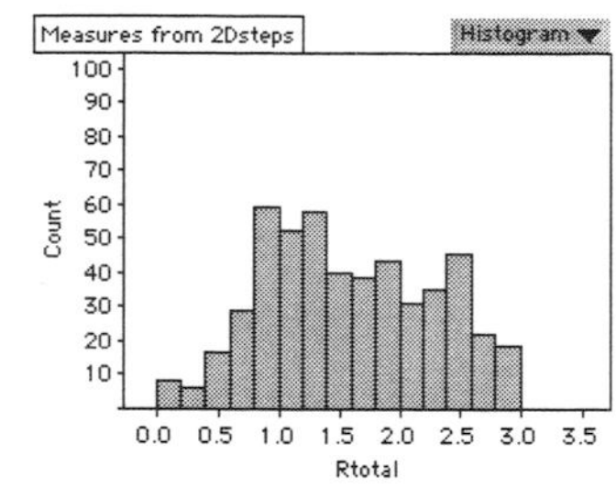

Notice the "density ring" at about $r = 1$, and that the whole character of the distribution has changed since we added the third step.

- Add a fourth step and collect measures. The density ring is gone!
- Add sixteen more steps for a total of twenty. Now see what it looks like.
- Finally, choose **Show Hidden Objects** from the **Display** menu. Now you can see a distribution where each coordinate is random normal. Note how the distribution of radii looks the same (even though the scales are different).

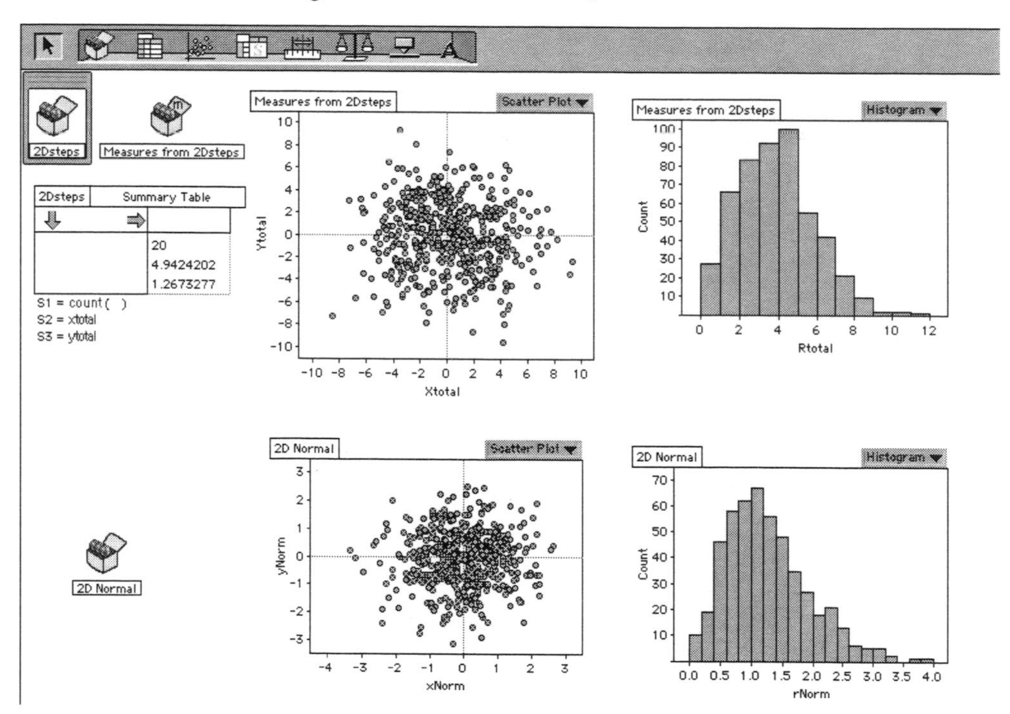

We haven't shown that the distribution is normal, but just as with the one-dimensional random walk—which is actually binomial—it *looks* normal after enough steps. Also, note how non-normal the distribution of *radii* is.

Standard Deviation, Standard Error, and Student's t

We already talked about standard deviation in the opening set of demos in this book, "Measures of Center and Spread" on page 21. And in "Standard Scores" on page 36, we used that spread to create the *z*-score—a way of describing a point in the context of its distribution, independent of how we measure the data.

This section takes these ideas further, using another, related measure of spread: standard error.

If you're learning statistics—for the first or for the *n*th time—it can be hard to stay really clear about what standard error is, and why you should care. Similarly, it's easy to treat the *t* statistic only as something that the computer makes for you in order to feed the black box of the *t* test.

These demos give you one way to make sense of these slippery concepts.

In this section, you will find:

"Standard Error and Standard Deviation" on page 66. This demo lets you experience these two statistics visually, and see how they change with sample size.

"What Is Standard Error, Really?" on page 68. Here we look at the standard error as the spread in the sampling distribution of the mean. We also look at that spread more quantitatively and, as we did in "How Random Walks Go as Root N" on page 51, see a root-*N* dependence.

"The Road to Student's t" on page 71. Now we add another layer of complexity: suppose we have a sample and want to figure out how far the sample's mean might be from that of the population. We discover that, *if we measure that distance in terms of the standard error*, the distribution we get is not normal. We need a new distribution: *t*.

"A Close Look at the t Statistic" on page 75. Here we use a fictitious sample of three points and see what happens to the *t* statistic as we drag the points around. We see how we can get a big *t* if we put all the points on one side (that's obvious—their mean should be far from the population value) and clump them together (that's not obvious, but it lowers the standard error, raising *t*).

web

Remember: If you see web in the side margin, that means that there's a solution on the web at the time of publication. See **http://www.eeps.com**.

Demo 16: Standard Error and Standard Deviation

Getting a feel for the difference between standard deviation and standard error

It's easy to get standard deviation and standard error confused. Not only do they both start with *standard*, they're closely related. This informal demo should help you develop some intuition about the difference; "What Is Standard Error, Really?" on page 68 goes into greater depth about standard error, and helps lead towards understanding the *t* statistic.

What To Do

- Open the file **SD and SE.ftm**. It should look like this:

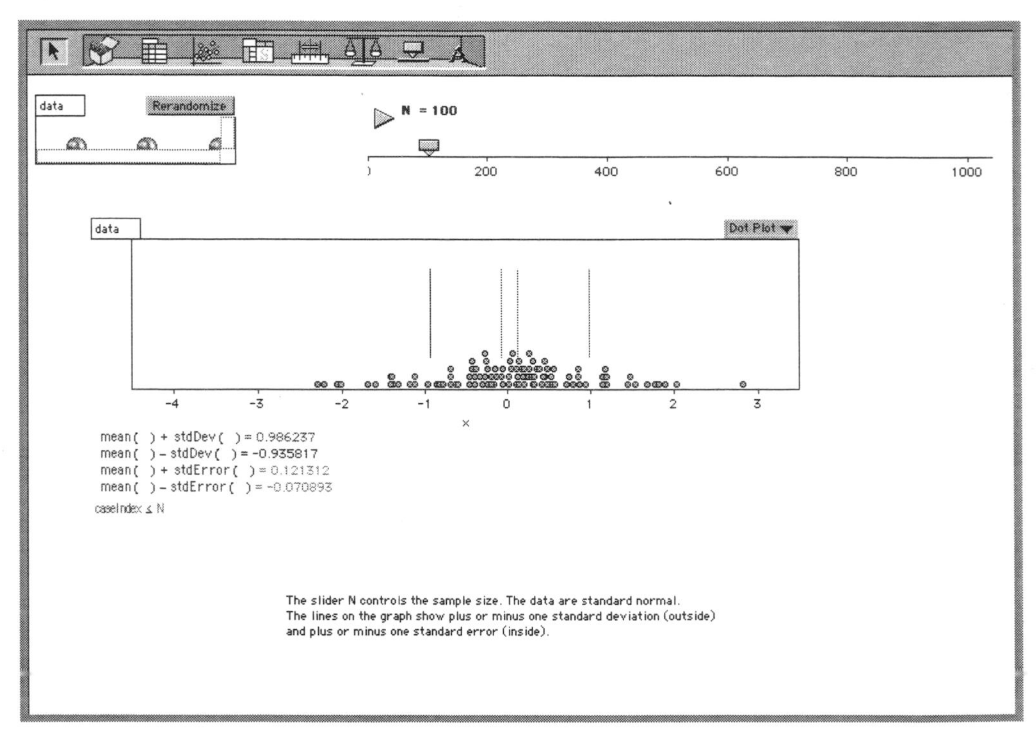

This relatively uncomplicated window contains a single collection, **data**, in the upper left. It has 1000 cases and one attribute, **x**. Fathom pulls these **x**-values randomly from a "standard" normal distribution—one with a mean of zero and a standard deviation of one. The graph shows the distribution of the first **N** cases, where **N** is controlled by the slider.

The graph also shows the positions and numerical values of the mean ± one standard *deviation* (the outside pair of lines) and the mean ± one standard *error* (the inside pair).

- Drag the slider **N** to the right. Note how the outside lines stay roughly in the same places—near plus or minus one.
- Do the same thing, paying attention to the positions of the inside lines—the ones that show how big the standard error is. Be sure to take **N** down well below 100.

You should have seen that, though it may jump around a little, the standard deviation (SD) interval remains roughly constant, but the standard error (SE) interval gets larger as **N** decreases.

What is the difference between these two? Let's see:

- Reset **N** to 100 (you can edit the number in the slider and press **enter**, **tab**, or **return**). Press the **Rerandomize** button in the collection (upper left). The points will change in the graph, and the lines will move.

- Repeat, noting what proportion of the time the SE interval encloses the true mean (0.00).
- Do the same thing, but with **N** = 400.

You should see that in both cases, the SE interval captures the true mean about 2/3 of the time. The SD interval, on the other hand, captures roughly 2/3 of the *data*, but almost always covers the true mean.

You can think of it this way: if you pick a point at random from the distribution, it is likely to be about 1 SD[1] from the mean. But if you pick a sample from the distribution and compute *its* mean, how far is *that* likely to be from the true mean? About 1 SE.

Extension

Let's study the SE interval more closely.

- Reset **N** to 100. Press the **Rerandomize** button.
- Repeat, noting the *width* of the SE interval.
- Record the full width of that interval at **N** = 25, 100, 400, and 900. Get several values at each **N**. (You could even enter these in a new Fathom collection with two attributes: **N** and **width**, and type in the values you recorded—or you could write a formula and collect measures.)
- Note how, even though the interval jumps around, its width decreases (and becomes more consistent) as **N** increases. In particular, note that the width is roughly $2/\sqrt{N}$ —that is, 0.4, 0.2, 0.1, and 0.067.

The root-*N* dependence of standard error is the same as that of the spread in random walks (as in "How Random Walks Go as Root N" on page 51). This is no coincidence. In both cases we're talking about how the spread of a sampling distribution decreases with the size of the sample. The random walk is about proportion; this one is all about a "regular" mean.

1. Depending what you mean by a likely distance; we're speaking informally here. If you really mean root-mean-square distance, it's 1.00 SD by definition. On the other hand, the mean absolute distance for normal data is actually about 0.8 SD.

Demo 17: What Is Standard Error, Really?

The connection between standard error and the sampling distribution of the mean • How the sample size connects standard deviation and standard error

The difference between standard deviation and standard error is important, and has everything to do with another concept, *sampling distributions.*

This demo gives you examples of all three of these close together so you can tease them apart. Here's the main idea: If you take samples of size *n* from a population, and take the mean of each sample, you can construct a sampling distribution—the distribution of those means. That distribution will be roughly normal (see "The Central Limit Theorem" on page 96 for restrictions). Its mean will be the mean of the population, and its standard deviation will be the standard deviation of the population *divided by the square root of n.*

What To Do

- Open the file **What is SE.ftm**. It will look something like this:

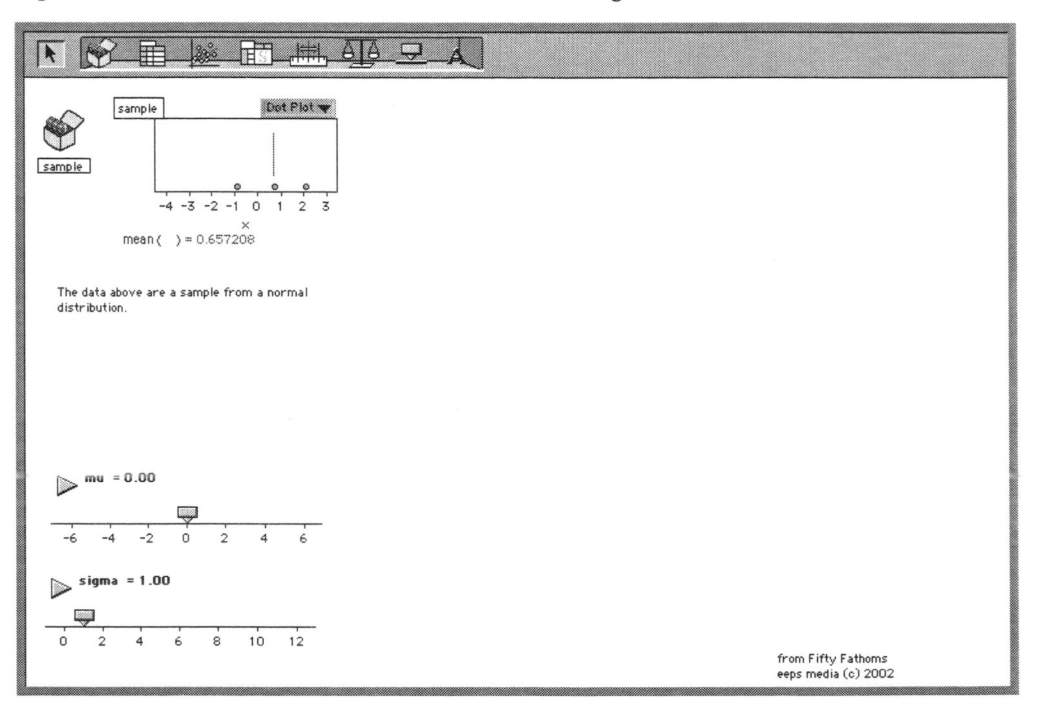

This window shows a collection called **sample**—the box in the upper left—with three cases. These are drawn from a normal distribution with mean **mu** and standard deviation **sigma**, controlled by the sliders. The little graph shows the values (**x**) of these three points.

The shortcut for **Rerandomize** is **clover-Y** on the Mac, **control-Y** in Windows.

- Click once on the **sample** collection to select it. Then choose **Rerandomize** from the **Analyze** menu, repeatedly. You should see the three points in the graph—and their mean—jump around.
- It would be great to collect those means. Let's do that: first choose **Show Hidden Objects** from the **Display** menu.

The newly revealed objects—the collection called **Measures from sample**, the summary table to its right, and the large graph—will show the results of our resamplings of the **sample** collection. The summary table will show the numerical mean and standard deviation of those resampled means. That is, **Measures from sample** is a *sampling distribution.*

- Press the **Collect More Measures** button in the measures collection. The **xbar** graph fills with data; each point is a different mean from those three **data** values.
- Note (in the summary table) that the standard deviation of this distribution of means is probably about 0.60.

Now we are going to investigate the spread of that sampling distribution as we change the number of cases in the **sample** collection.

- Click once on the **sample** collection to select it.
- Choose **New Cases...** from the **Data** menu. Add one case. You should see the new point appear in the graph.

Note: the right-hand graph *does not change*, because we have not collected the measures. We have to do that ourselves:

- Again press the **Collect More Measures** button in the measures collection. The graph updates. Note that the standard deviation of this distribution has changed. Repeat a few times. You should get a standard deviation near 0.50.
- Now that we see what this is a distribution of, let's change the display. In the popup menu in the corner of the graph, choose **Histogram**. The graph will look something like the one in the illustration.

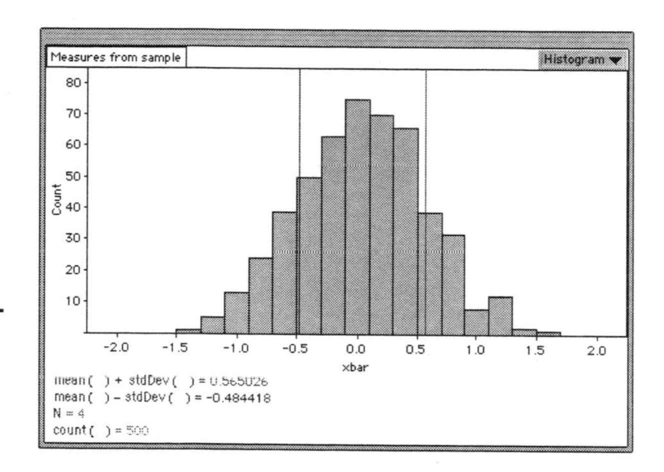

- Add 21 cases to **sample** for a total of 25. **Collect More Measures** again. See how the standard deviation decreases.
- Add 75 cases for a total of 100. Again, **Collect More Measures**.

With $n = 4$ we get a standard deviation of about 1/2, with 25 we get about 1/5, and with 100 about 1/10. One conjecture is that n is the square of the denominator; in other words, the standard deviation of the sampling distribution is the population standard deviation (**sigma**) divided by the square root of n. Let's see:

- Click on the summary table once to select it.
- Choose **Add Formula** from the **Summary** menu. The formula editor opens.

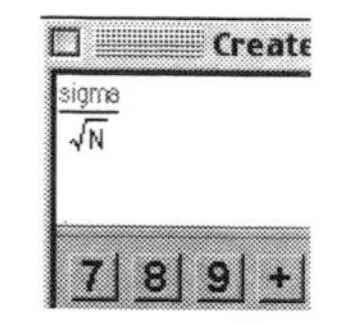

- Enter **sigma**/$\sqrt{N}$ as shown. (**N** is defined in the measures collection. Use the button for square root.) Press **OK** to close the editor.
- The newly-computed quantity appears in the table. Compare it to the standard deviation above it.
- See if this near-equality holds true in other cases: change the number of cases, the population mean **mu**, and the population spread **sigma**.

This quantity—the standard deviation divided by $\sqrt{N}$—is called the *standard error*, and the whole point is that it's the standard deviation of the sampling distribution. That may sound useless, but let's put it another way: the SD is a measure of how far a given data value is likely to be from the mean of the sample. The SE is a measure of how far that mean of the sample is likely to be from the mean of the *population*. That's the SE: the SD of the sampling distribution.

Note: Usually, when we compute standard error, we do so having only the results of a single sample; we divide the *sample* SD by $\sqrt{N}$ to get the SE. Here, we know the population SD, **sigma**, which we never do in real life.

Questions

1 The sampling distribution of the mean looks more or less normal. Is it? How would you know?

web

2 What standard deviation do you expect in the sampling distribution if there are 16 cases in each sample, and the original standard deviation is 1.0? Try it and see if you're right.

Harder Stuff

web

3 With the sampling distribution as a histogram, make Fathom draw the relevant normal curve so you can compare. (You may want to use a density scale on the graph—in the **Graph** menu, look under **Scale**; you can what it looks like in the next demo.)

4 Suppose we collected 1,000 measures instead of only 500. Explain what difference that would make in this demo and why.

Extensions

- Assess the normality of the sampling distribution. Change the histogram to a normal quantile plot (use the popup menu in the corner of the graph) and see if the data are in a straight line.
- Suppose the original data were not drawn from a normal distribution but, for example, from a uniform distribution. Would the sampling distribution of the mean still be normal? Change the formula for **x** to be **random( )** and see what changes.
- Suppose the data were binary? Use **randomPick(–1, 1)** for **x** to see what happens.

Demo 18: The Road to Student's t

Using standard error as the scale for measuring how far a sample mean is from the true mean • How these quantities are not normally distributed; in fact they follow a t distribution

Where have the last two demos led us? Why do we care about this difference between standard deviation and standard error? Here's what we have seen:

- Standard deviation measures the spread in the sample, and therefore reflects the spread in the population.
- The standard error of the mean measures the spread of a sampling distribution, and is therefore a measure of how well we know the mean of the population if we have only a sample.

This second point sounds obvious, but it isn't, and is worth pondering. If your eyes fall shut when you hear "sampling distribution" (even though we tell you it's important), remember it this way:

The SE—the standard deviation divided by $\sqrt{N}$—is the standard deviation of the sampling distribution. The standard deviation is a measure of how far a given data value is likely to be from the mean. So if you take a sample and compute its mean, how far is that mean likely to be from the true mean of the population? The standard error.

That's what we said in "What Is Standard Error, Really?" on page 68. Now we look more deeply and uncover one of the Great Subtleties of Statistics.

What To Do

Open the file **Road to t.ftm**. It will look something like this:

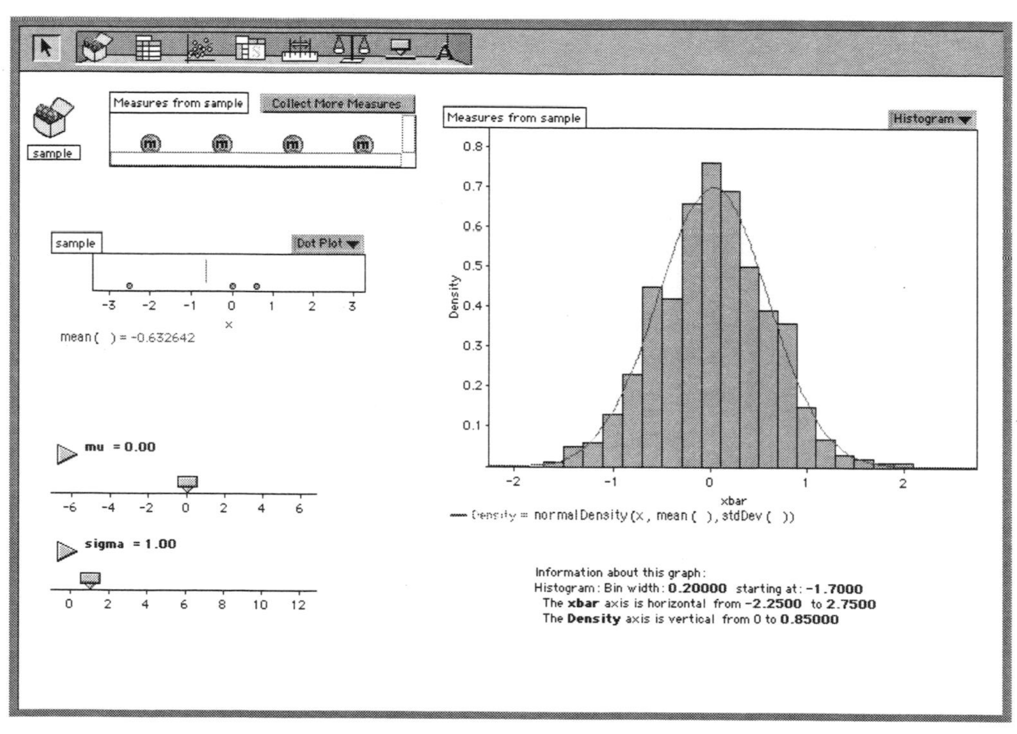

This is based on the file **What Is SE.ftm**, if it looks familiar. We see the **sample** collection (three cases, normally distributed, sliders for **mu** and **sigma**, plus a graph) and its derived **Measures from sample** collection. The histogram shows the sampling distribution of 500 means (called **xbar**), with the relevant[1] normal curve superimposed. Looks pretty normal.

---

1. We're using the *data's* mean and SD to define a normal curve. We could have used slider parameters in our formula, e.g., **normalDensity(x, mu, sigma/**$\sqrt{\text{num\_cases}}$ **)**, but using the data works well here.

- Just to show that it wasn't a fluke, press the **Collect More Measures** button on the top-middle collection, and generate a new set of 500 **xbar**s.

We can predict the shape of the sampling distribution if we sample repeatedly. We take the standard deviation, divide by $\sqrt{N}$, and use that as the standard deviation for our normal distribution. If we know about the Central Limit Theorem (page 96), we even know it's normal.

A Longer Explanation Than Usual

Here comes the hard part, conceptually: in real life, we only get the one sample, and we don't know what the true mean or standard deviation are. When we get that sample, we'd like to know how far that sample is likely to be from the mean.

We're tempted to say, "Let's take the sample standard deviation and divide it by $\sqrt{N}$ to get a standard error. Now we can reverse the traditional logic of the standard error, like this: Just as there's a 95% chance that the sample mean will be within two standard errors of the true mean, we have 95% confidence that the true mean is within two standard errors of the sample mean."

We'd be wrong. That symmetrical logic breaks down, as we shall see. The demo will go like this: we'll pretend for a moment we don't know "the truth" when we draw the sample. We pull three numbers **x** from the distribution and calculate the mean, **xbar**. We want to get some idea how far that is likely to be from the true mean.

Here is the crux of the matter: it's not fair to calculate that distance in absolute units. If we really don't know how far it is to the true mean, that answer should not depend on the scale—for example, the system of units—we measure with. But we *can* express the distance *in terms of the spread of our sample* (or spread of the *population*, but we don't know that). We could use the standard deviation of our sample as a unit of measurement, but since we know that the distance from the sample mean to the true mean scales like standard error (not standard deviation) let's use that. So far, we've matched the tempting logic above.

Then imagine that we're told the truth, so we can see how far off we really are. Now, instead of plotting the errors in absolute units, we'll make the experiment independent of the particular scale, and calculate the errors in *units of sample standard errors*. And plot them. Let's go...

Note: this trick, scaling the data this way, uses *dimensionless* quantities. If the original measurements were in centimeters, say, this will give us an analysis that has no units at all. That way it will scale properly even when we change systems of measurement.

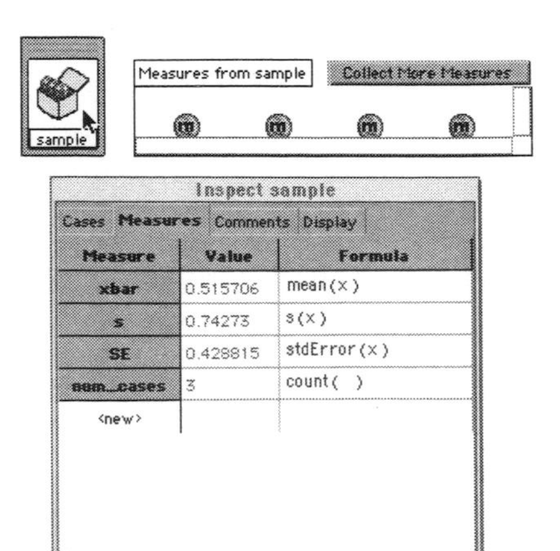

- First, we have to calculate this dimensionless "difference." Double-click the **sample** collection (not **Measures from sample**) to open its inspector.
- Click the **Measures** tab to open that panel.
- Make a **<new>** measure; call it **error**.
- Double-click **error**'s formula box (to the right) to open the formula editor.
- Enter **xbar / SE** and press **OK** to close the editor (these attributes are both defined in this same panel, so it's OK to use their names). Note: we assumed **mu = 0**, so **xbar** *is* the deviation.
- Close the inspector to save screen space.

We've defined **error** to be a dimensionless number—the distance from the sample mean (**xbar**) to the true mean but in units of standard errors. Now let's see its distribution:

- Press the **Collect More Measures** button on the measures collection. Fathom collects 500 new **xbar**s. Where are the **error**s? We need to tell Fathom to plot them.
- Double-click the **Measures from sample** collection (the open box, not its name) to open its inspector.
- Click the **Cases** tab to open that panel.
- Drag the name of the attribute **error** from the inspector to the horizontal axis of the graph, replacing **xbar**. Shazam! The graph updates, and we see the distribution of this new attribute.
- It looks bizarre, like the top graph at right. We need to rescale the axes and change the bin width. You can do it by hand, or double-click the graph to make "ControlText" appear below the graph, as shown. Use a bin width of 0.5, and a range of at least –5 to +5.

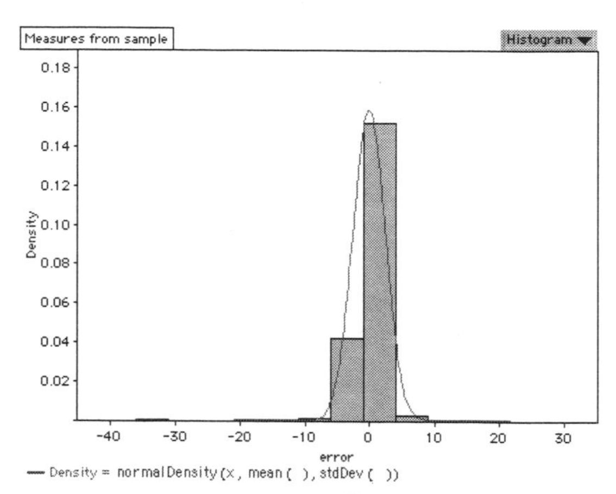

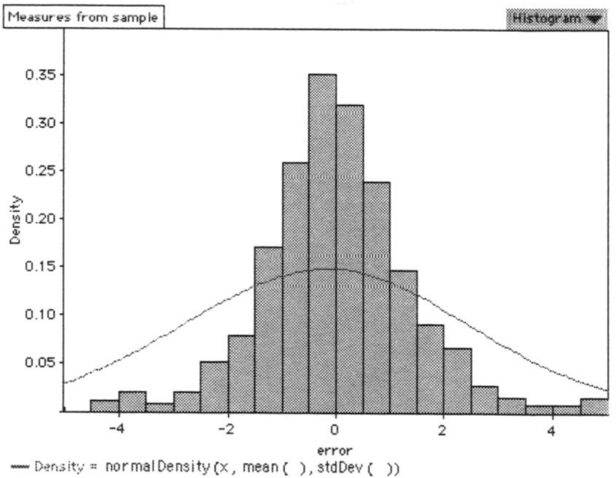

Information about this graph:
Histogram: Bin width: **0.50000** starting at: **-33.000**
The **error** axis is horizontal from **-5.0000** to **5.0000**
The **Density** axis is vertical from 0 to **0.39931**

The graph does not match the curve—a normal curve with the same mean and standard deviation as the data.[1] So there must be some really far-out-there points to make the standard deviation so large. At any rate, we have found that *even though the xbars are normally distributed, these error quantities are not* (at least with $n = 3$).

If it's not normal, what is it? Let's see:

- Click the graph once to select it.
- Choose **Plot Function** from the **Graph** menu. The formula editor opens.
- Enter **tDensity(x, N – 1)**. Press **OK** to close the editor and show the graph.

It should match! This distribution was Gosset's great discovery, what we now call Student's t distribution. What we called *error* we could just as well call t. Just to emphasize:

$$t = \frac{(\bar{x} - x_0)}{SE},$$

1. We could simply have plotted **normalDensity(x)** to get a standard normal curve, which looks (at first glance) much closer to this distribution. It is just as unsuitable, however, because of the data out in the tails. With the curve we have drawn, the difference between the histogram and the function is more dramatic.

where x_0 is a value you're comparing the sample to. Of course, in this case we needed to know the true mean, but the symmetrical logic that failed us above now holds water. The point is that if you measure these differences in terms of standard errors, you don't need to know the population spread—because it's all done using the scale of the *sample* standard error.

Measuring in terms of spread also helps us clarify this important distinction:

- A difference measured in units of sample standard deviation (effect size) is an indication of how *meaningful* the difference is. That is, it says how much distributions overlap, or how far apart they are in terms of their spreads.
- A difference measured in units of standard error of the mean (Student's *t*) is an indication of how statistically *significant* the difference is. That is, it helps us understand how likely it is that the difference occurred by chance.

Extensions

These extensions are worth some time and thought if you're studying the *t* distribution, *t*-tests, or confidence intervals based on *t*.

- Change the **error** histogram to a normal quantile plot. What does that graph tell you? (Note that when you change back, you may have to remind Fathom to display a *density* histogram. Choose **Scale>Density** from the **Graph** menu.)
- Change the value of **sigma** (and re-collect measures) to see that the **error** distribution (i.e., *t*) stays the same even when the population spread changes.
- Change the value of **mu** (and re-collect measures) to see that things get weirder. That's because our definition of error assumes that the true population mean is zero (it should have been **(xbar – mu)/SE**). What we get is a *noncentral t* distribution.
- Add additional cases to the **sample** collection. See how the shapes of things change as sample size changes.

Harder Stuff

1. Look back at the histogram of **error**. Maybe it *is* normal, but the normal curve just needs to be rescaled to fit the graph. After all, it looks about the right shape—it's just that the SD is too big. Try making a slider for a parameter to rescale the SD, and see what you find. (One suggestion for a formula: **normalDensity(x, mean( ), stdDev( )/K)**, where **K** is the slider.) It fits pretty well—what is it about the graph that says it does not fit well enough?
2. One way of characterizing this new distribution is that it's kind of normal except that it has longer tails. Why is that? What do you suppose the tail-samples tend to have in common? What about the ones in the hump? ("A Close Look at the t Statistic" on page 75 might help you look at this.)
3. [web] Why didn't we just measure the difference from the true mean in standard *deviations* instead of standard *errors*? It would be dimensionless—independent of the original units—and in a scale that depended only on the sample we got. Try it, see what happens, and explain your results.
4. We made this statement above: "Just as there's a 95% chance that the sample mean will be within two standard errors of the true mean, we have 95% confidence that the true mean is within two standard errors of the sample mean." We claimed that this statement (in its context) was wrong. Or at least misleading. What's wrong with it?

Demo 19: A Close Look at the t Statistic

How sample mean, standard deviation, t, and P interrelate • How they depend on the values of individual points in a sample

In this demo, we'll test a sample of three points. We'll use the *t* test of the mean to see if we can distinguish that mean from zero. In fact, we won't really care about the results of the test. Rather, we'll be looking at data from many tests to learn how the center and spread of the data relate to the *t* statistic and to its associated *P*-value.

What To Do

- Open the file **Close Look at Student's t.ftm**. It should look something like this:

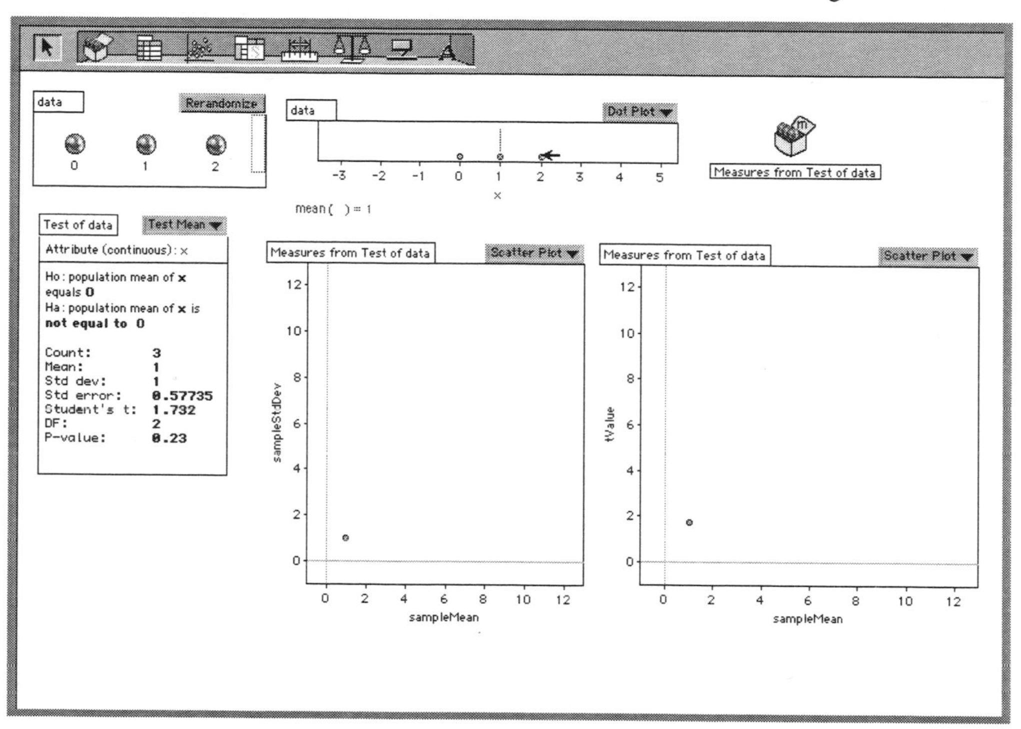

This document has the original data at the upper left, the *t* test below it, a graph of the data top center, and two big graphs in the middle that display the results of the test. As we change the data, Fathom will add more points to the graph as fast as it can. (Fathom collects these test results in the collection upper right; the graphs are of data from that collection.)

Observe how the two graphs display the sample mean, standard deviation, and *t*-value in the test that you can see. (If the test is unfamiliar and you'd like a more complete explanation, select it and un-check **Verbose** from the **Test** menu; you'll need to shrink it later to make enough screen space to see both the test and the graphs.)

Under Windows, a dialog box may flash as you drag. If this happens, try to ignore it.

- In the short, wide graph at the top, grab the right-hand point—the one at a value of 2, as shown in the illustration—and drag it to the right. See how the mean (**sampleMean**) and standard deviation (**sampleStdDev**) both increase in the two graphs below, but that **tValue** decreases.
- Now drag it all the way to the left, watching the two graphs. Be sure you get points near the maximum or minimum in both graphs. Return the point to near 2 when you're done. (**Undo** is perfect for this.)

Questions

web

1. Where do you put the "2" point to make the standard deviation a minimum? Why?
2. When you moved the point to the right, why did t decrease? Isn't the mean getting farther from zero?
3. Is the place where t is a maximum different from the point where the standard deviation is a minimum?

Onward

- Now, with the "2" point back where it belongs, drag the "1" point back and forth.
- Finally, put the "1" point back and drag the "0" point. The right-hand graph will look something like the one in the illustration.

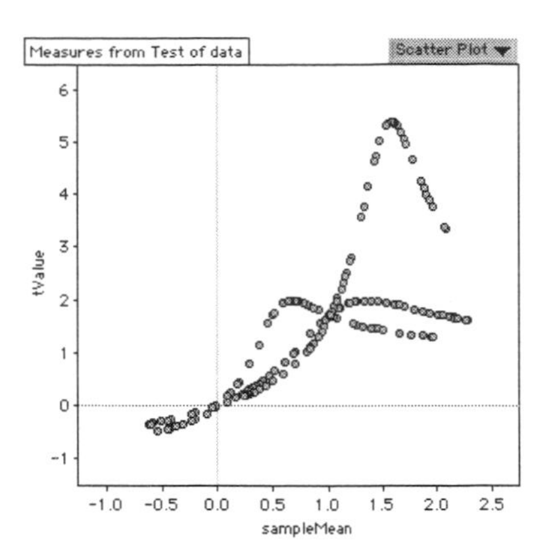

More Questions

web

4. Which point do you have to move to get the smallest standard deviation? Why? (Trick question alert.)
5. Which point do you have to move to get the largest t?
6. Explain the previous answer in terms of the situation; that is, since it's a test to see if the mean is zero, explain (ideally without formulas) why moving *that* point makes the null hypothesis the *least* believable.

Extension

Suppose we want to explore the *P*-values. In which of the tests we have done would we reject the null hypothesis $\mu = 0$ at the 5% level?

- Double-click the highest point on the **sampleMean** vs. **tValue** graph—the point with the largest value for t. The collection's inspector opens, and you should see the value for p (called **pValue**), which is about 0.03. So we would reject that one.
- Drag the name of the attribute **pValue** into the *middle* of that scatter plot. The points color, and a legend appears at the bottom. The purplest points are for low p, and would be rejected.
- Let's get a more accurate picture, using the other (left-hand) graph. First, drag the attribute **tValue** to the horizontal axis of the left graph, replacing **sampleMean**. The graph will look like the one in the illustration.
- Take a moment to appreciate this graph! It shows how the largest values of t for the situations we have looked at have small standard deviations.

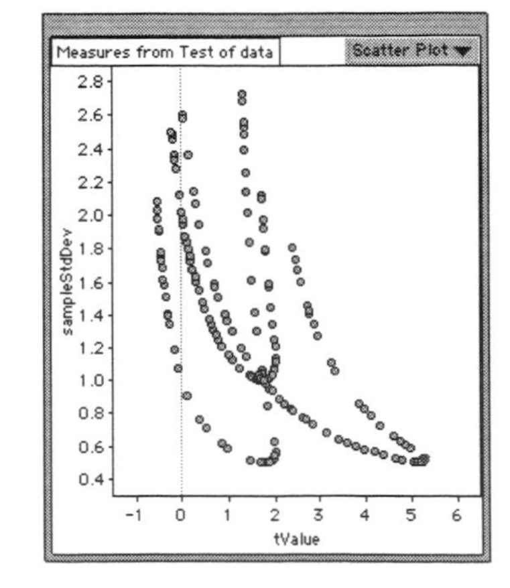

Note that this also suggests an explanation for why the t distribution is different from the normal: the points in the long tails of the t distribution are from samples whose means are offset from the true mean, but happened to have a small spread. This often happens with three points when all three are on one side of the mean (it happens 1/4 of the time, after all). They will naturally have a smaller SD than the average sample. A small SD means a small SE, and a larger *t* statistic. Similarly, a sample with a large SD will be more likely to straddle the mean because it's so fat. But not only is the sample mean closer to the true mean, the *t* statistic will be pushed closer still because the difference is measured in units of SEs—and there are fewer SEs between the mean and the population value because of the large spread.

Of course, the more points there are, the less likely you are to get a sample SD that's very extreme—and the closer the distribution gets to normal.

Now drag **pValue** to the vertical axis of that graph, replacing **sampleStdDev**. This graph shows how p depends on t. Large t—positive or negative—and you get a small p. Which is what you need to reject the null hypothesis.

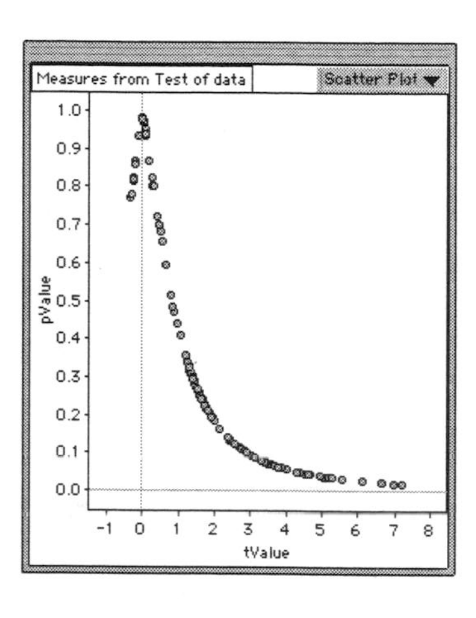

Harder Stuff

1. Zoom into this last graph to figure out what values of t correspond to values of p less than 0.05. Confirm that they're the same as those in a critical-values table.
2. Add an additional point to the sample: select one of the left-hand gold balls, **Copy** it (**Edit** menu) and **Paste** it (**Edit** menu again). The new point will appear in the dot plot, and the *t*-test will report a sample size of 4 instead of 3. Drag these points. Explain what happens on the t vs. P graph—why isn't it the same as before? How is it different?
3. web — As in the first task, zoom in to see where the critical values for t are. There are now two—one for a sample of three, one for four. Confirm that the one for four is smaller than the one for three and explain why.
4. Select all of the points in the data graph (the short, wide one at the top) by dragging a marquee (dashed rectangle) around them. Then drag them all at once. Describe what happens in the (now messy) graph of **tValue** vs. **sampleMean**, and explain it.

Sampling Distributions

Sampling distributions show up all over the place in statistics, and the idea of sampling distributions occurs throughout this book. Many of the relevant demos are in other places; but here, in this section, is a collection of additional ideas held together by this common thread.

In this section, you will find:

"The Distribution of Sample Proportions" on page 80. Here you get to see how this distribution depends on the population proportion and the sample size. Of course, we don't see only a *theoretical* (in this case, binomial) distribution—we see real samples, with all their attendant variability.

"Sampling Distributions and Sample Size" on page 85. Now we again take sampling distributions of the mean, but this time we compare these distributions for different sample sizes. As we discovered in "What Is Standard Error, Really?" on page 68, the distribution gets skinnier as sample size goes up.

"How the Width of the Sampling Distribution Depends on N" on page 88. Here we look quantitatively at the spread of the sampling distribution. We use the IQR and discover the inverse root-N dependence related to the one we discovered in "How Random Walks Go as Root N" on page 51.

"Does n – 1 Really Work in the SD?" on page 91. We can use sampling distribution—in this case, of the sample standard deviation—to assess whether a statistic is an unbiased estimator. We discover that it is not. See also "Sample Variance: Why the Denominator is n – 1" on page 178.

"The Central Limit Theorem" on page 96. You change the population and see that the sampling distributions of most—but not all—statistics are usually—but not always—pretty normal.

More Peripheral Issues

We have also included three demos in this section that aren't really part of the direct inferential road, but have more to do with making good estimates:

"Adding Uniform Random Variables" on page 81. If you sample from two uniform distributions and add, what's the distribution of the result? (Triangular.)

"How Errors Add" on page 83. Suppose the two things you add are measurements with a normally-distributed error. What's the error in the sum? We see that we get the Pythagorean sum of the errors.

"German Tanks" on page 93. In the context of this famous problem, we look at estimators and their sampling distributions in order to see if they're unbiased or not.

Demo 20: The Distribution of Sample Proportions

How sample size and population proportion affect the distribution

This simple demo lets you see quickly how the population proportion and the size of the sample affect the distribution of sample proportions.

What To Do

- Open the file **Dist of Sample Props.ftm**. It looks something like this:

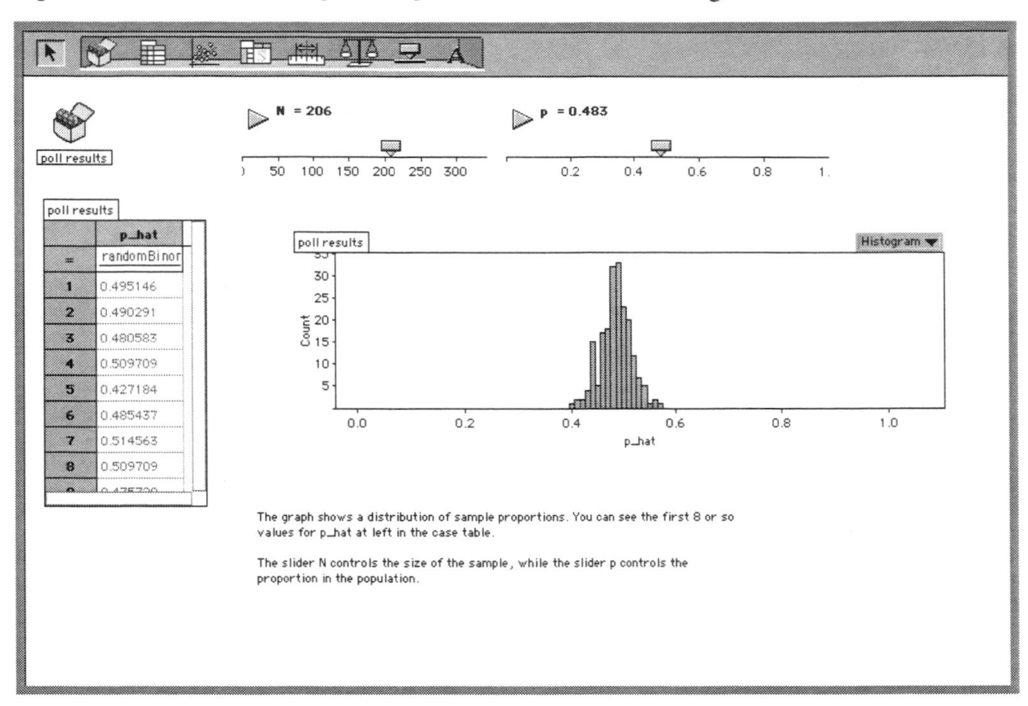

Suppose you're conducting polls about something where the respondents say simply yes or no. You control the number of people in each poll and the population proportion of "yes" by the sliders **N** and **p**, respectively. The graph of **p_hat** shows the distribution of sample proportions from 200 polls.

- Play with the sliders and see what happens. Answer the questions.

Note: due to strange behavior on Fathom's part, avoid setting **p** to exactly **0.00**. If you do, the graph will go wacky. Our friend **Undo** does not help us here, but fear not; re-choose **Histogram** from the popup and rescale.

Questions

1. What's the main thing that happens as you move **p**?
2. What's the main thing that happens as you move **N**?
3. When is the distribution asymmetrical?
4. (web) If you poll a sample of 100 people from a town, and in reality, 55% of the town wants new sewers, it is possible that less than half your sample will say they want new sewers. About how likely is that?
5. If you poll 50 people, and in reality 40% of a population says "yes," what's the range of sample proportions you're likely to get?

Demo 21: Adding Uniform Random Variables

What happens when you add two uniform random variables • How that corresponds to adding two dice

This demo gives a brief look into distributions of uniform random variables—what they look like, how they vary, and what happens when you add them. We'll start by looking at a familiar, though discrete, uniform distribution: that of a fair die. We know what happens when you add two dice; does the same thing happen when you add uniformly-distributed random numbers?

What To Do

- Open the file **Adding Uniform.ftm**. It will look something like this:

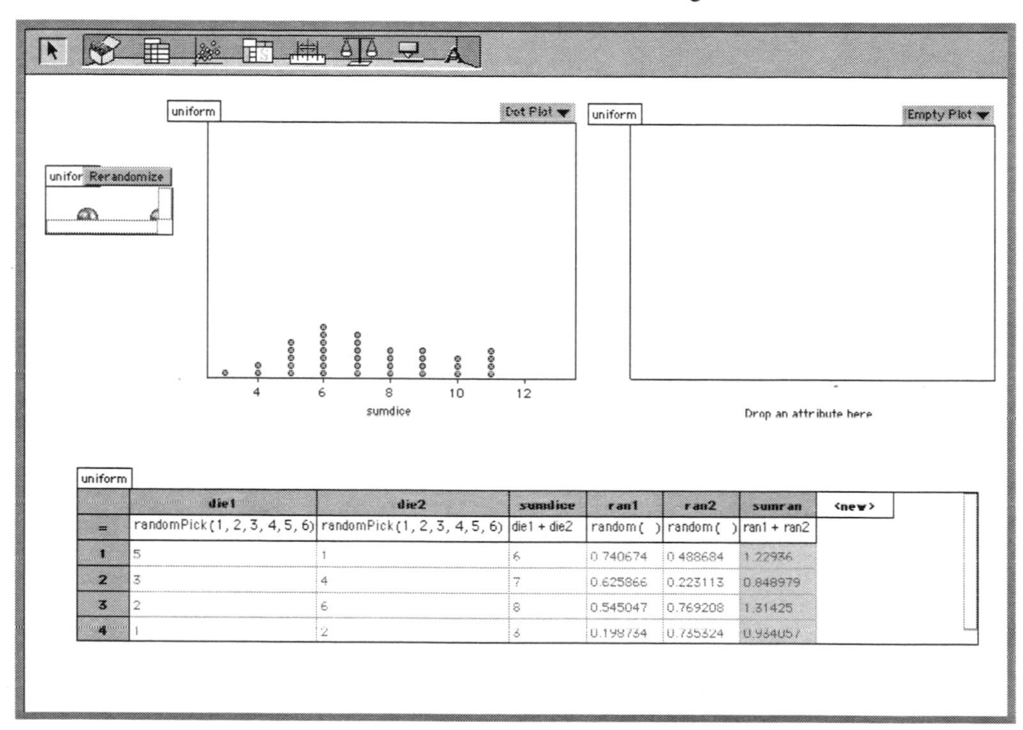

The case table at the bottom shows the data and, in its top row, the formulas for each of the attributes (columns). You can see that **die1** and **die2** are fair dice (read their formula) and that **sumdice** is their sum. The left graph shows the distribution of 36 sums of dice. The attributes **ran1** and **ran2** are uniform random numbers in the range 0 to 1; **sumran** is their sum.

- Press the **Rerandomize** button in the collection near the upper left. The random numbers and the graph change. Repeat to get a sense of how varied the distributions can be.

- Drag the column head—the attribute name—**ran1** to the horizontal axis of the right-hand graph. Fathom makes a dot plot of the random numbers. You'll see a graph like the illustration.

- Press **Rerandomize** again a few times; see how these "uniform" random numbers often bunch up or leave gaps.

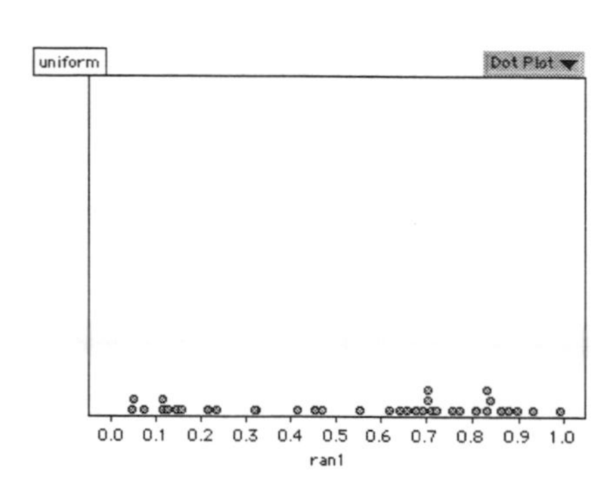

- Drag the attribute **sumran** to that same horizontal axis, replacing **ran1**. Except for a change of scale and a slight bunching towards the center, it probably looks about the same.

The dice bunch up, why not the uniform numbers? Let's add more cases.

- Choose **New Cases...** from the **Data** menu. Add 64 cases for a total of 100. At this point your window will look something like this:

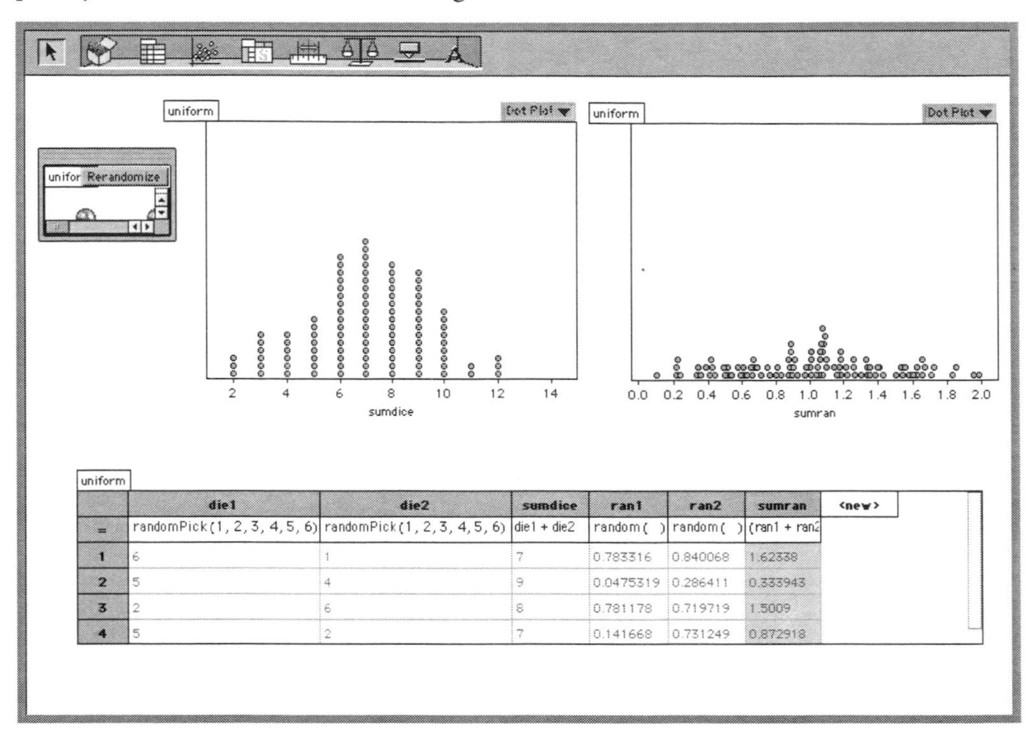

You may need to alter the vertical scales for comparison.

- Change both graphs to histograms (using the popup menus in the graphs themselves). Now the graphs probably look a lot more similar.
- Put **die1** and **ran1** on the horizontal axes, replacing the **sum**s. See how flat the histograms look (i.e., not very). Press **Rerandomize** to see how varied they can be.
- Add more cases! With anything selected, choose **New Cases...** again from the **Data** menu. This time, add 900 cases, for a total of 1,000. See how the graphs change? They look flatter. **Rerandomize** to check.
- Now replace **die1** and **ran1** with **sumdice** and **sumran**, respectively. You should see the characteristic triangular shape that you get when you analyze the dice theoretically.

It's funny how different representations can make things look. In a dot plot, the uniform numbers and the dice numbers look rather different, but in a histogram, they look a lot more the same. The point is that you get the same sort of result with uniform continuous random variables that you do with their discrete cousins, die rolls. Since it may be easier to analyze die rolls at first, you can use that analogy to help you understand uniform distributions.

Extension

Extend the demo to add three or four or more random numbers. What happens to the shape of the distribution?

Demo 22: How Errors Add

Basic error analysis • If two quantities each have some measurement error, finding the error in their sum

It is a curious thing that when you add two (independent, normally-distributed) random variables, their means add in the usual way, but you take the Pythagorean sum of their standard deviations to get the standard deviation of the sum. That is,

$$\sigma_{tot} = \sqrt{\sigma_1^{\ 2} + \sigma_2^{\ 2}}$$

This demo lets you see that happen.

What To Do

- Open the file **Adding Errors.ftm**. It will look something like this:

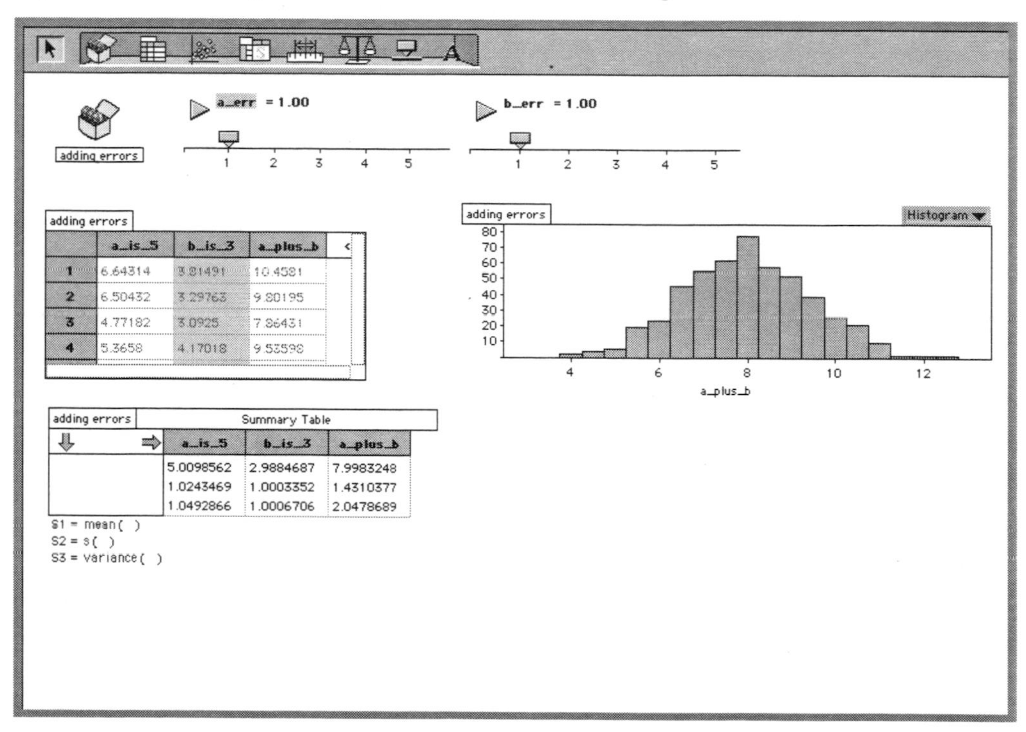

The collection **adding errors** contains the data you can see in the case table below it: values for **a_is_5**, **b_is_3**, and their sum, **a_plus_b**. The attribute **a_is_5** has the value 5, plus a (possibly negative) random error whose standard deviation is controlled by the slider **a_error**. Analogously for **b_is_3**. Below the case table is a summary table where you can see the means, standard deviations, and variances of those three attributes. The distribution of the sum appears in the graph.

The shortcut for **Rerandomize** is **clover-Y** on the Mac or **control-Y** in Windows.

- Choose **Rerandomize** from the **Analyze** menu. Note in the top row of the summary table that the means seem to add (the means of **a** and **b**—5 and 3—add to 8, which is about the mean of the sum), but that the standard deviations—the middle row of the summary table—do not ($1 + 1 \neq 1.4$).
- Move the sliders **a_error** and **b_error** to see how they affect the graph and the numbers in the summary table.
- Verify that nothing you do changes the means very much.

- Verify that if **a_error** is large compared to **b_error**, then the standard deviation of **a_is_5** is pretty much the same as the standard deviation of **a_plus_b**.
- Verify that if **a_error** is about the same as **b_error**, the standard deviation of **a_plus_b** is about the same as 1.4 times **a_error**.
- Verify that, though the standard deviations do not add in the obvious way, the *variances*—the numbers in the bottom row of the summary table—seem to.
- Verify that if you set one error to 3 and the other error to 4, the error of the sum is about 5. (This is just like a 3-4-5 triangle. The variances should be about 9, 16, and 25—but with quite a bit of slop.)

If that's true, then we can add the mean squares of the errors to get the mean square error of the sum. That is,

$$SD_a^2 + SD_b^2 = SD_{total}^2, \text{ or}$$

$$SD_{total} = \sqrt{SD_a^2 + SD_b^2}$$

Let's look at the data again, recording what we find out and comparing it to that conjecture.

- Return the two error sliders to 1.0.
- Choose **Show Hidden Objects** from the **Display** menu. A measures collection and a graph appear.
- Double-click the measures collection to open its inspector. Tell it to **Recollect measures when source changes** by clicking in that checkbox. Close the inspector.
- Now drag the **a_error** slider and watch what happens. Be sure to go both above and below 1. The graph will update to show you how the standard deviation of the sum **SDsum** changes as a function of the standard deviation of **a_is_5**, which is **SDa**.

Let's put the function in.

- Click once on the scatter plot to select it, then choose **Plot Function** from the **Graph** menu. The formula editor opens.
- Enter the formula $\sqrt{x^2 + \text{b\_err}^2}$, as shown in the illustration. Press **OK** to close the editor. Notice how well the function models the points.
- Set **b_error** to 3, and then leave it there as you change **a_error** again.

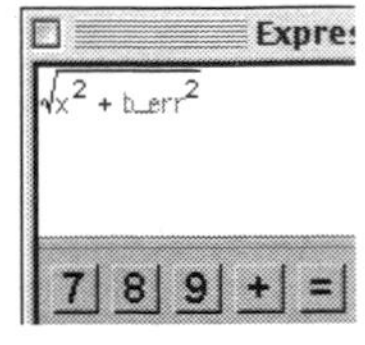

Harder Stuff

web 1. It looks as if the function is pretty straight if you get far from the *y*-axis. Find the equation for that line and explain why it has to be that way—using reasoning about errors.

web 2. The function looks flat as you get close to the *y*-axis. Explain that too.

Theory Corner

You can read about why variances add in the section "Variance" on page 173.

Demo 23: Sampling Distributions and Sample Size

How sampling distributions (of the mean) get narrower as you increase sample size

It makes intuitive sense that you get a better idea of a statistic—a mean, say, or a proportion—if you use a larger sample to estimate that statistic. But you seldom get to see the effect of sample size in a single graph. In this demo, you get to see it: we will build sampling distributions of the mean for samples of different sizes; and as usual, you will control the mean and standard deviation of the population using sliders.

Note: Why do we assume the population is normally distributed? In general, it's not. But for this demo, it's a convenience: you could use any distribution, as we do in "The Central Limit Theorem" on page 96, but that requires extra machinery on the screen that is beside the point of this demo. With a distribution that we have described with a random-number formula instead of by actually sampling, we need only the two collections—the sample and its measures—instead of three (population, sample, and measures).

What To Do

- Open the file **Sampling Dists Sample Size.ftm**. It will look something like this:

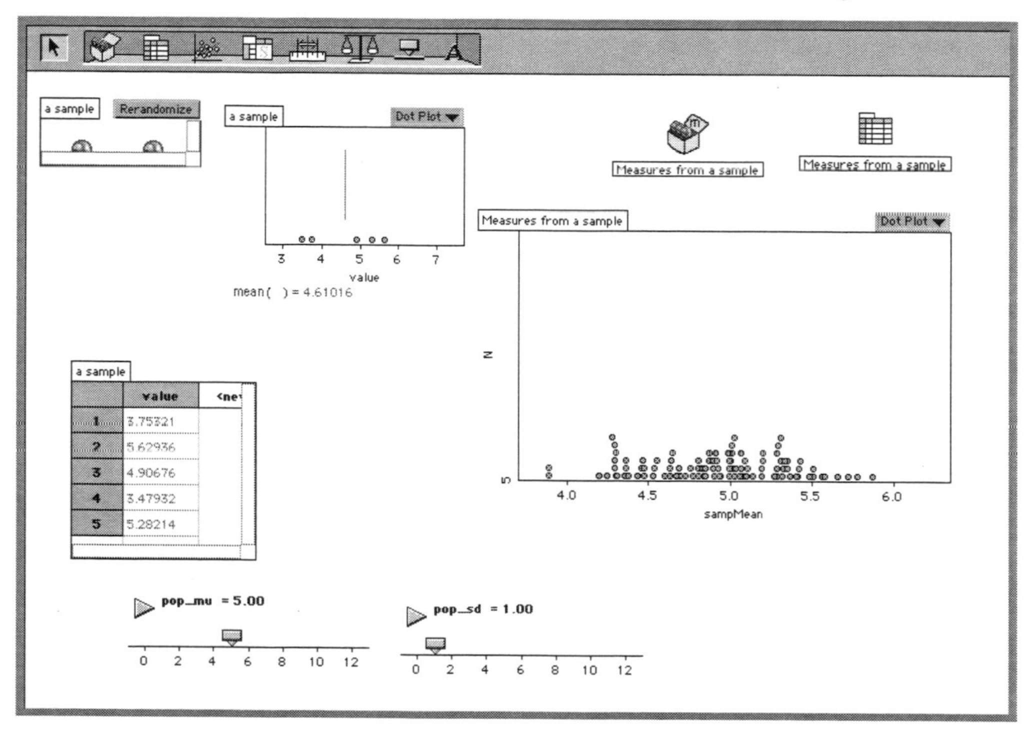

The collection on the left, called **a sample**, has five points in it (though some points may be off the graph). The values are random, drawn from a normal distribution where the mean is 5.0 and the standard deviation is 1.0. On the right is a collection of measures from that sample—at this point, 100 means of that sample. That is, we chose 100 samples of five, computed their means, and plotted them. The vertical axis—you can see the "5" at the bottom—is the sample size.

Before we explore sample size, we should look briefly at the left-hand collection (**a sample**) alone.

- Press the **Rerandomize** button in the left-hand collection. Note how the graph and table change. You can see the mean in the graph. See how it changes from press to press. (Here people often wonder if there is an automated way to collect all of those means.

That's what *measures* are for; in fact, that's exactly what the right-hand graph is: 100 of those means.)

- Verify informally that the variation of the means you get is roughly the same as the variation of the means (named **sampMean**) in the right-hand graph.
- Click the collection **a sample** once to select it. Then choose **New Cases...** from the **Data** menu. Give the collection—our sample—15 more cases (for a total of 20).
- Click the collection **Measures from a sample** once to select it. Then choose **Collect More Measures** from the **Analyze** menu.

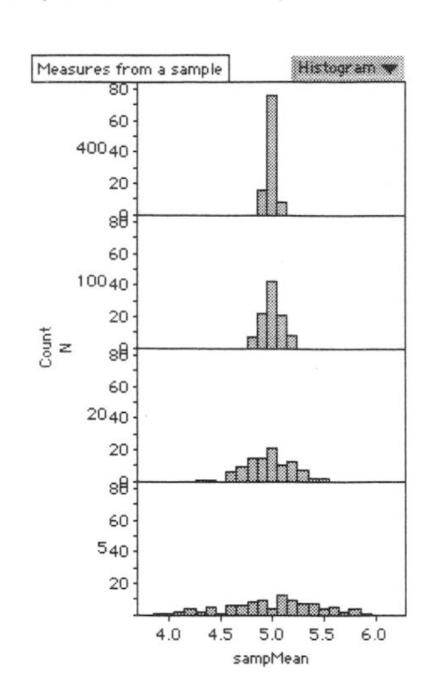

Notice how the right-hand graph updates. It collected 100 more samples—this time of size 20—computed their means, and plotted them. The graph is set up to separate the different sample sizes.

- Repeat the last two steps, adding 80 more cases (for a total of 100); collecting measures; and then adding 300 more cases (for a total of 400), and collecting measures again. There are so many dots now, they overlap. (That's what the red dots mean.)
- Change the dot plots to histograms by choosing **Histogram** from the popup menu in the graph. Your graph should show four histograms.

To Slow Down the Process

This business of collecting measures can seem mysterious. It might help to slow it down and see what happens in a little more detail. Fathom lets you do that. Here's how:

- Double-click the measures collection to bring up its inspector.
- Click the **Collect Measures** tab to bring up its panel. This is where you control how the measures are collected.
- Change the panel to turn animation on, as shown. You may also want to reduce the number of measures you collect (for the sake of speed) from 100 down to 25.

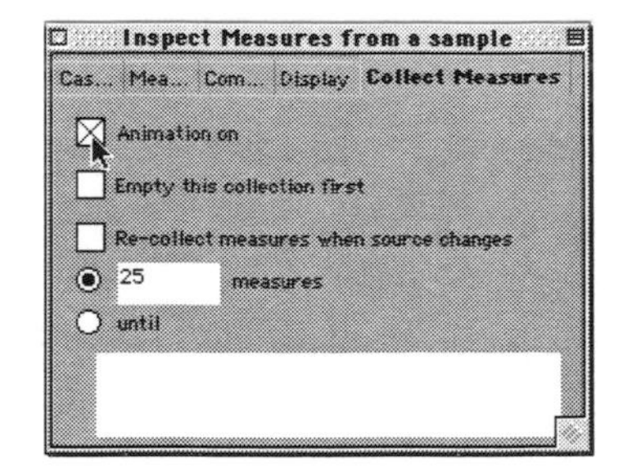

- Close the inspector and proceed as before: select the **Measures from a sample** collection, then choose **Collect More Measures** from the **Analyze** menu. You'll see the new cases appear (on the graph of the sample means) one at a time instead of all at once.

Questions

1. Why does the mean change when you press **Rerandomize**? Isn't the mean just 5.0?
2. web — What do you notice about the distributions of the sample means when you increase the sample size?

Extension

You don't have to collect sample *means*. You can make sampling distributions of any statistic. Try sample *medians*, for example. This will be easy, because these collections have actually been collecting sample medians (and sample maxima) all along:

- Double-click the measures collection in order to open its inspector.
- Click the **Cases** tab. You'll see a window like the one in the illustration.
- Drag the attribute name **sampMedian** to the graph to replace **sampMean**.

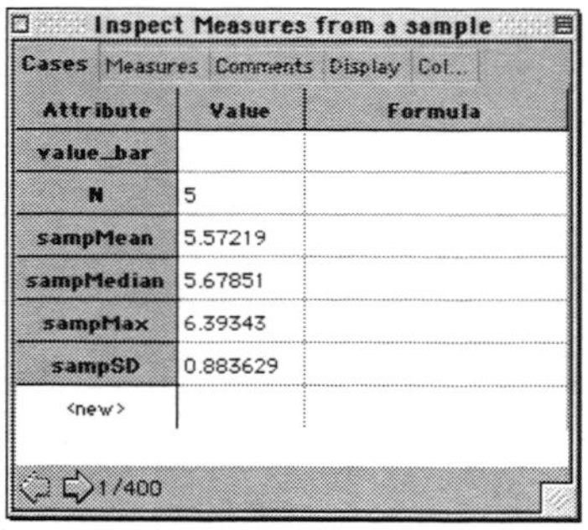

Inspect Measures from a sample

Cases | Measures | Comments | Display | Col...

Attribute	Value	Formula
value_bar		
N	5	
sampMean	5.57219	
sampMedian	5.67851	
sampMax	6.39343	
sampSD	0.883629	
<new>		

1/400

More Questions

3 How does the median graph look the same as (or different from) the graph for means?

4 In general, the bigger the sample size in the first collection, the narrower the distribution in the second, measures collection. Explain why that happens in plain language.

Another Extension

web

Do the same as in the previous extension, but for *standard deviation* (use **sampSD**) instead of median. In this case, the distribution gets narrower, as before, but something else happens as well. What is it?

We discuss this in excruciating detail in "Does n – 1 Really Work in the SD?" on page 91.

Note

The distributions on the right, the ones in the measures collection, are called *sampling distributions*. In many books, they care only about sampling distributions of the mean, which is what we did first. But you can make sampling distributions of any statistic you like. In Fathom, make it a measure in your original collection: open the inspector by double-clicking the collection; click the **Measures** tab to go to that panel; and define the measure there by giving it a name and formula. Then, when you collect measures, it will be an ordinary, graphable attribute in the measures collection. (We prepared the collections in this demo specially to make mean, median, and max easy to study.)

The Central Limit Theorem is all about sampling distributions becoming more normal as *n* increases. But not all sampling distributions do that. If you make a sampling distribution of **max( )**, for example, it will not gradually become normal—or even converge to a particular value. (You can see it as we did with the median; just use **sampMax** instead of **sampMean** or **sampMedian**.)

Demo 24: How the Width of the Sampling Distribution Depends on N

How the width (as measured by IQR) of a sampling distribution of the mean is inversely proportional to the square root of the sample size

This demo follows on the heels of "Sampling Distributions and Sample Size" on page 85.

By the conclusion of that demo, we had collected sampling distributions of the mean repeatedly for different-sized samples. We saw that the distribution of those sample means got thinner as n increased. The question is, how *much* thinner do these distributions get?

This is important because the width of these distributions parallels the width of the corresponding confidence interval. That is, if we're estimating a statistic such as the mean (or the median, or the standard deviation) we need to predict how good our estimate is likely to be. We know that it will be better the bigger the sample is—*how much* better is what we'll find out. In practical terms, larger samples cost more. Does the extra accuracy legitimize the extra cost?

In this demo, we'll essentially start with the end of the previous demo and extend it.

What To Do

- Open the file **Width of Dist Depends on N.ftm**. It should look like this:

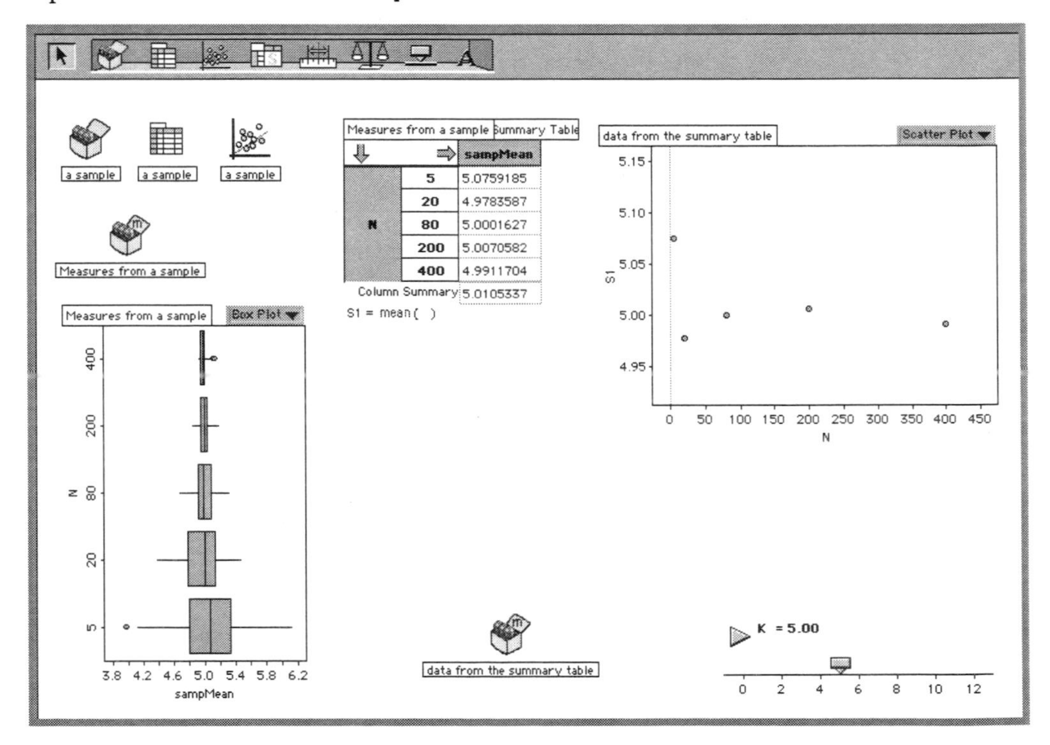

We have "iconified" the objects for the original sample in the upper left, and removed the controls for the mean and standard deviation of the sample—they're fixed at 5.0 and 1.0, respectively. We also moved the "measures" plot to the lower left and changed it from a histogram to a box plot. (You can change it back using the menu in the graph itself, if you wish.) Each box plot displays the means for 100 samples.

The new elements are a *summary table*, currently listing the means of the 100 sample means for each of the different sample sizes; a collection—a box of balls—representing the data in the table itself; and a graph of that data, currently showing the means of the sample means as a function of sample size. As you can see, they're all near 5.0, which is what we would expect.

Now we have to compute and display the spread.

- Click the summary table once to select it. A **Summary** menu appears in the menu bar.
- Choose **Add Formula** from the **Summary** menu. The formula editor appears.
- Enter **iqr( )** (with nothing inside the parentheses). Close the editor with **OK**. A new number—the interquartile range (IQR)—appears in every cell of the summary table, as shown in the illustration. You can see by the legend that the second element—called **S2**—is the interquartile range. Note how the IQR gets smaller as **N** gets larger.

Measures from a sample / Summary Table

	N	sampMean
N	5	5.0759185 0.54626529
	20	4.9783587 0.37034813
	80	5.0001627 0.18221526
	200	5.0070582 0.10402531
	400	4.9911704 0.059883105
Column Summary		5.0105337 0.16931581

S1 = mean( )
S2 = iqr( )

- Now we want to put the IQR on the graph. Double-click the collection—the box at the bottom of the window—called **data from the summary table**. That opens its *inspector*.

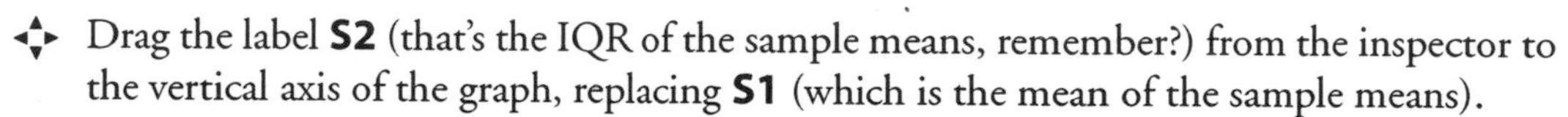

- Drag the label **S2** (that's the IQR of the sample means, remember?) from the inspector to the vertical axis of the graph, replacing **S1** (which is the mean of the sample means).

You should see a graph like the one in the illustration. We can see that, indeed, the IQR (**S2**) decreases as n increases, but what is the functional form of the decrease? Suppose we think that an inverse relationship is a good model for the data, that is,

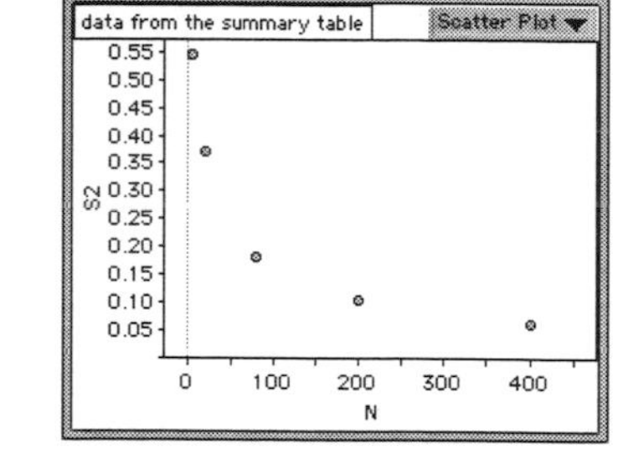

$$iqr(n) = \frac{K}{n},$$

where K is a constant (it's also the name of the slider that we'll use). We'll make this mistake—and then we'll fix it.

- Click the graph to select it. Then choose **Plot Function** from the **Graph** menu. The formula editor appears.
- Enter **K/n**. (Note that you do not enter the "S2 =" part of the function—just the right-hand side.) Close the editor with **OK**. A function appears.
- Use the slider named **K** to make the function fit as well as you can. You'll see that no value of **K** makes the curve fit well—it must be the wrong function.
- Double-click the function (**S2 = K / n**) at the bottom of the graph to edit the function.
- Change it to be $K/\sqrt{n}$, as shown in the illustration. You can use the square-root key in the editor to get the radical sign. Close the editor with **OK**.

- Again, use the slider **K** to fit the curve. It doesn't fit perfectly, but it does a lot better than the "straight" inverse did.

Extensions

Let's see what happens when we use a different measure of spread. We'll use sample standard deviation instead of interquartile range.

- Close the inspector.
- At the bottom of the summary table, double-click where the formula reads **S2 = iqr( )**. In the formula editor, enter **s( )**, which is the sample standard deviation. Now the graph of **S2** plots **s( )** instead of **iqr( )**. Use the slider to fit the function to the data.

- You get a different value for **K**. Larger or smaller? Why?
- As before, we have also collected **sampMedian** and **sampMax** in the measures collection. You can put those in the summary table and explore how their means and spreads change with sample size.

Harder Stuff

1. Use the **percentile( )** function to compute a 90% band instead of IQR, and re-collect the measures to get the graph. Does this fit a curve better? Worse? About the same?
2. What does **K** really mean (especially when you use SD instead of IQR)? Why does the point at **n=5** not fit the curve as well?
3. web — Suppose you didn't think of using the square root in the denominator—you just saw that the inverse function didn't work. What could you do, in exploring the data, to "discover" the square root relationship?

The Point

The point of all this is to become more familiar with the properties of sampling distributions. Here, we see that for the mean at least, the spread of its sampling distribution decreases more or less as the square root of the sample size—no matter what we use to measure spread.

It also helps to see that we can treat results from statistical simulations as data, and we can treat them the same way as we would if we were modeling a physical phenomenon with a function.

Theory Corner

We discuss why this works from a theoretical standpoint in "The Distribution of the Sample Mean" on page 174.

Demo 25: Does n – 1 Really Work in the SD?

Unbiased estimators • How the familiar formula for sample standard deviation is not unbiased • Why we should care about variance

Amazingly, no. Not exactly. Here's a common misconception. See if it sounds familiar:

Suppose you draw a sample from a population, and measure some continuous attribute for every case. Estimating the mean of the population is easy: the mean of the sample is an unbiased estimate of the mean. Estimating the standard deviation of the population is easy, too: you use the standard deviation of the sample—almost. Instead of n in the denominator, though, inside the square root sign, you use $n - 1$. You can't use n because that SD is not an unbiased estimator of the population standard deviation, whereas the one computed with $n - 1$ is. That is,

$$SD_{fromsample} = \sqrt{\sum \frac{(x - \bar{x})^2}{n - 1}}$$

What's not to like? Where's the problem? It's true that we're supposed to use $n - 1$, but *it is not true that the SD so calculated is an unbiased estimator of the population SD*. If this is old news, O Reader, more power to you. You don't need this. But it was a big surprise to me, the author, so I thought I'd share it with anyone else who might have labored under the same misconception.

What To Do

Open the file **Estimating SD from a Sample.ftm**. It should look something like this:

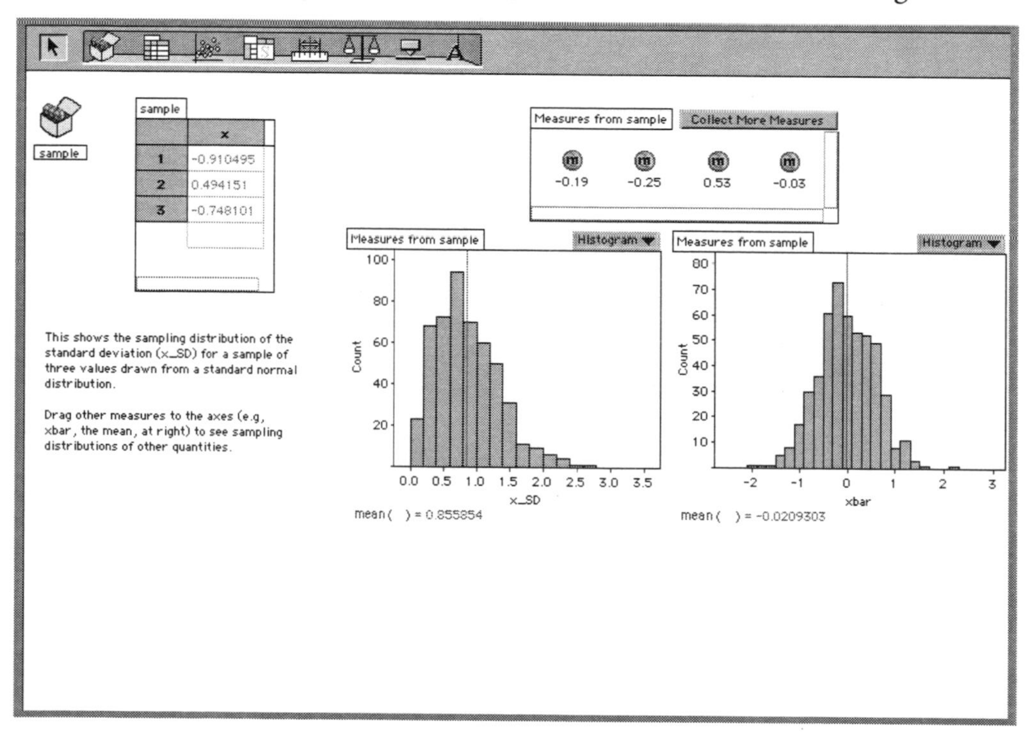

On the left you can see three values (**sample**) drawn from a normal distribution with a mean of 0 and a standard deviation of 1. The left-hand graph is a distribution of standard deviations (**x_SD**) calculated from 500 different samples, and the mean of those 500 SDs. On the right is a sampling distribution of the mean, just like the sampling distributions we have seen, for example in "The Road to Student's t" on page 71. But the graph on the left is a sampling distribution of the SD, rather than a sampling distribution of the mean.

Notice that the mean of **x_SD** is less than 1.0, which is the "true" value. Also, the distribution is skewed. What's up?

First, let's see whether the mean is really as low as it looks.

- Click **Collect More Measures** in the **Measures from sample** collection. The graph will update with the SDs from 500 new samples. It should still be low.

Next, let's see what the "plain" SD (made with n instead of $n - 1$) would look like graphed.

- Double-click the Measures collection to open its inspector. Drag **x_popSD** from the **Cases** panel to the horizontal axis of the right-hand graph, replacing **xbar**. Its mean should be even lower!

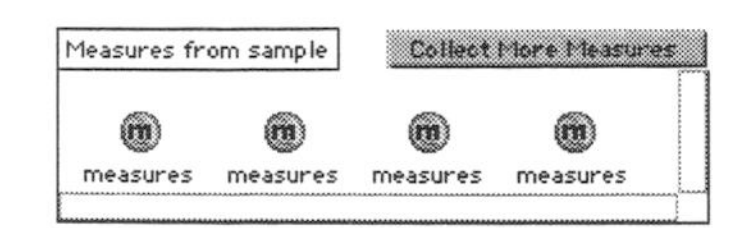

Okay, let's see what's really going on:

- Double-click the original sample collection (the one with three cases) to open its inspector. Click the **Measures** tab to bring up that panel.
- Make a new measure by clicking in the **<new>** cell. Enter **x_Var** (for "x variance").
- Double-click **x_Var**'s formula cell (two boxes to the right of the name) to open the formula editor.

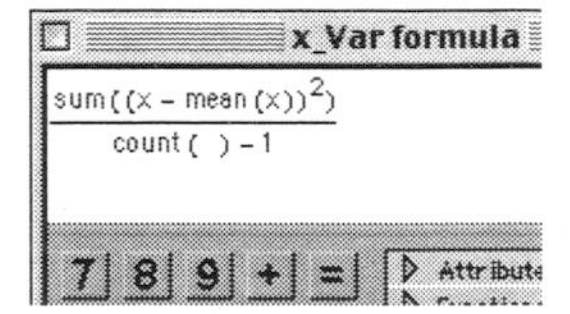

If you have trouble entering the formula, just type the characters in order (with all parentheses). Or you could simply use **Variance(x)**, since **Variance** is a built-in function—but writing out the formula may be more appropriate here.

- Enter **sum((x − mean(x))^2)/(count( ) − 1)**, as shown. (That's the variance of x, using $n - 1$ because this is a sample.)
- Close the formula editor with **OK**. Verify that Fathom computed a variance (it should be the square of the first standard deviation). Then close the inspector to clean up the screen. (The inspector for the measures collection should remain open.)
- Again, click **Collect More Measures** on the measures collection. This time Fathom collected variances as well.
- Drag the new attribute, **x_Var**, from the inspector of the measures collection to the horizontal axis of the right-hand graph, replacing **x_popSD**.
- Click on one of the bars to show where these points are in the **x_SD** graph.
- Resample the variances (i.e., click **Collect More Measures**) a few times, until you're convinced that, unlike the standard deviations, they really do center around 1.0.

That is, the sample standard deviation (the one with $n - 1$) is not an unbiased estimate of the population standard deviation. *But the sample variance is an unbiased estimate of the population variance*, even though the variance is just the square of the SD.

This helps answer the question of why we should ever care about the variance. Most important, though, in my opinion, is that it shows how a simple transformation of a variable can radically change how a distribution looks.

Questions

1. Why do you suppose we did this with a sample size of three as opposed to, say, 100?
2. (web) How could we have predicted that the mean of the "plain" SD distribution would be smaller than the first one we tried?
3. How is it possible that the distribution of the SD can look so different from that of its square, the variance?

Demo 26: German Tanks

Unbiased estimators • Evaluating estimators from their sampling distributions • Even among unbiased estimators, some are better than others

What do we mean by an unbiased estimator? In this demo, we'll see at least two of them, as applied to the famous German Tanks problem. The idea is that German tanks have serial numbers, sequential, starting at one. (Historically, the numbers were on particular tank *parts*, not the whole tanks.) You want to estimate how many tanks there are altogether. You capture a small number of tanks (four, in this case) and read their serial numbers. Assuming they are a random sample, what do you do with the numbers to get a good estimate of the total?

The complete set of tanks is the population. We get a sample of four, and we want to estimate N, the size of the population. It's a strange problem.

Part of the problem is deciding what we mean by a good estimate. For now, we'll ask for an *unbiased* one. An *unbiased estimator* is a procedure (e.g., a formula) that produces an estimate with a special property: if you run this procedure many times, the *mean* of the estimates will be equal to the true quantity you're looking for. Of course, in practice, you only use it once. But in this demo, we'll use a strategy you should be familiar with: Knowing the population, we run our procedure on it repeatedly to see how well the procedure performs. In this case, we'll know in advance that $N = 1000$.

What To Do

- Open the file **Tanks.ftm**. It will look something like this:

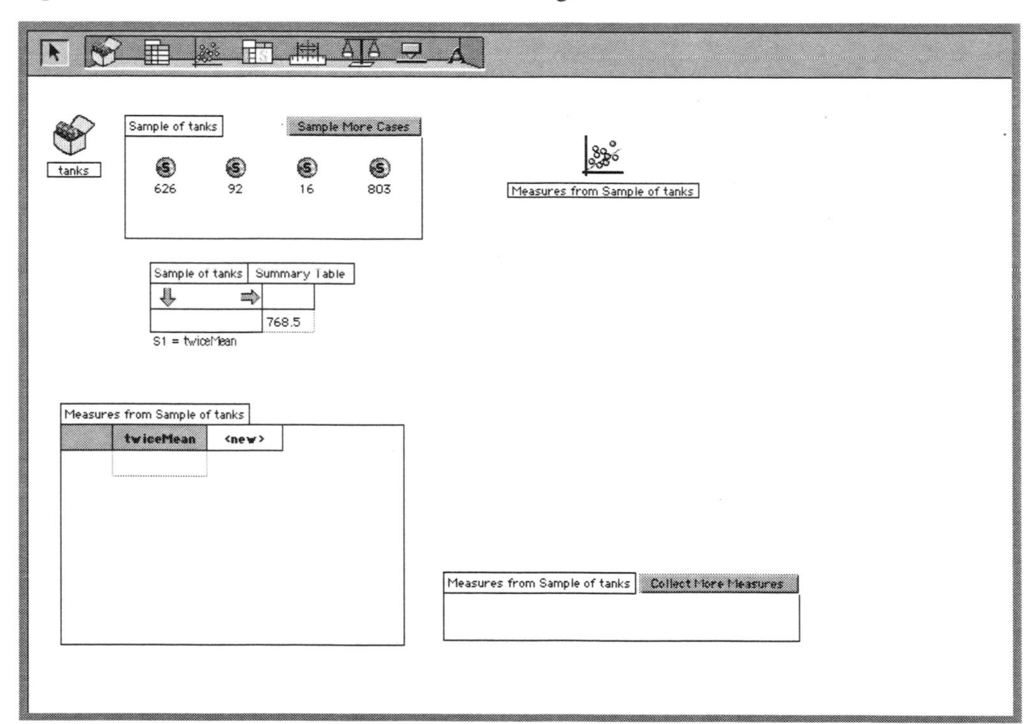

In the upper left is the original **tanks** collection. It contains 1000 cases, each of which has a single attribute, **serial**, that runs from 1 to 1000. Next is **Sample of tanks**, which is open (you can see the gold balls). It's a sample of four cases from the **tanks** collection. Below it, in a summary table, you can see an estimate of the population based on that sample, called **twiceMean**. Below those are a measures collection (**Measures from Sample of tanks**) and its case table, both empty. At the upper right is an iconified graph.

- Press the **Sample More Cases** button. A new set of four cases shows up in the collection, and our estimate changes down below. Repeat this a few times. How good is the estimate?
- Let's have Fathom collect those estimates automatically. Press **Collect More Measures** in the bottom collection. A column of numbers—200 values of **twiceMean** from 200 different samples—appears in the bottom table.
- Click on the graph icon on the right to select it; then drag its lower-right corner to expand the graph. You should see a graph like the one in the illustration.

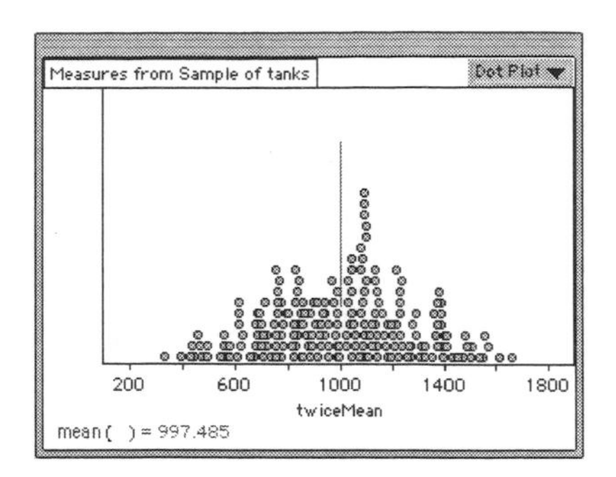

Note that the mean of these 200 estimates is close to the true number of tanks—1000—but that the individual values have quite a bit of spread. It appears that this measure is an unbiased estimator (or close to one); remember, however, that when you really do this to figure out the number of tanks, you only get one sample of four—only one of those two hundred points—and you don't know which one.

Questions

1. The measure **twiceMean** is just what it sounds like: twice the mean of the four serial numbers in our sample. Why is that a sensible estimate of the largest serial number?
2. What are the largest and smallest possible values for **twiceMean**? (web)

Onward!

Let's put in a different estimator of the total number of tanks—one called **Partition**. The idea is that if we pick four numbers and they're spaced out uniformly over the range, they could divide the population into five equal portions. If that were true, the largest serial number would be 5/4 of the *maximum* number in the sample. This strange reasoning yields a surprisingly good estimator.[1] (To skip the next set of steps, simply open the file **Tanks2.ftm**.)

- First we have to make the measure. Double-click the sample collection (the one with four balls) to open its inspector. Be sure the **Measures** panel is showing (click the **Measures** tab if you need to).
- Click in **<new>** to make a new measure; enter **Partition** and press **enter**.
- Double-click its formula box (to the far right of the name) to open the formula editor.
- Enter the formula **max(serial)*(count( ) + 1)/count( )**, as shown in the illustration. Close the formula editor with **OK**. Then close the inspector to save screen space.

If you just type the characters—including all parentheses—in order, the formula will work.

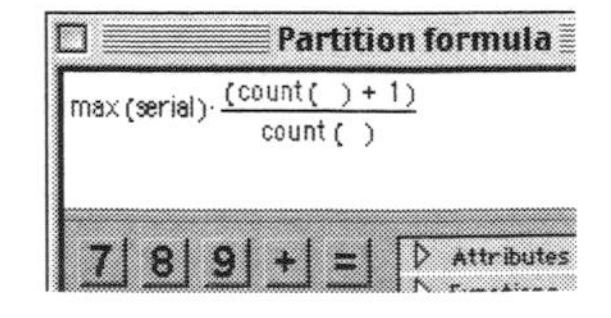

- Let's put the new measure in the table. Click the summary table to select it, then choose **Add Formula** from the **Summary** menu. The formula editor opens.
- Enter **Partition** and press **OK** to close the editor. The new value appears.

Note: This is the place you'll be if you open the file **Tanks2.ftm**. Also note that we used **count( )** in the fraction rather than just entering 5/4; this will keep the calculation correct in case you decide to change the sample size.

---

1. Though not unbiased. It's systematically high by 1—too small a difference here to matter.

- Press **Sample More Cases** (the top button) a few times to see the two values in the table. See whether you think that the new measure, **Partition**, is a good estimator.
- Press **Collect More Measures** in the bottom collection to do so automatically. Now there are two columns in the table—200 values of each measure.
- Drag the column head for **Partition**—the name itself—to the axis of the graph, replacing **twiceMean**. You should see a graph like the one in the illustration. What does it tell you?

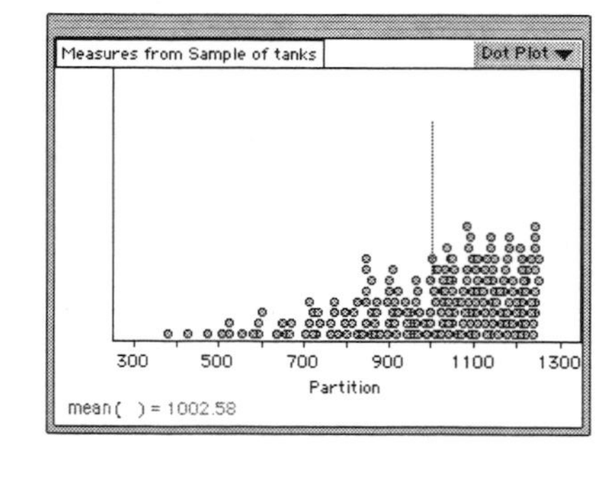

It seems that, though the distribution is very different from the one we got with **twiceMean**, **Partition** is also close to being an unbiased estimator, since its mean is close to the truth.

- Let's compare the two. Drag a new graph off the shelf and put it below the one we already have.
- Drag the name **twiceMean** to the horizontal axis of the new graph.
- Choose **Plot Value** from the **Graph** menu and enter **mean( )**. Close the formula editor with **OK**.

See "Rescaling Graph Axes" on page 18 for help adjusting the scales.

- Adjust the horizontal scales of the two graphs to match.
- Look at box plots as well as the default dot plots.

Even though we're pretty sure any estimate is not exactly correct, we can say something about whether we would rather have it high or low, and by how much. It may even be that a *biased* estimator would be better, depending on the circumstances.

Questions

1. What are the largest and smallest possible values for **Partition**?
2. Which distribution has the greatest range?
3. Which has the largest median?
4. Which estimator would you rather use to judge how many tanks there were? (Think about the consequences of overestimating or underestimating the number of tanks.)

Harder Stuff

5. Add additional estimators and assess them. Your own are best, but here are two to consider: twice the *range* of data (formula: **2 * (max(serial) – min(serial)**) and five (or **(count( ) + 1)**) times the minimum.

web

6. Rerun the simulation with a sample of ten tanks instead of only four. How do the distributions of the estimators change? To change that sample size—to capture more tanks—open the inspector for the **Sample of tanks** (opened) collection, and change the number sampled from 4 to 10.
7. It looks as if **Partition** is a better estimator than **twiceMean**. But **twiceMean** uses all of the data, whereas **Partition** uses only one point—the maximum. Just try and explain how that can be.
8. You could argue that it would be better to overestimate the number of tanks. Invent a situation where it would be better to *underestimate* the number of something that had serial numbers. What estimator would you use in that case?

Demo 27: The Central Limit Theorem

A demo of the CLT • How sampling distributions usually look normal • Cases where they do not

In a wide variety of situations, if you take a sample and calculate a statistic—for example, the mean of some quantity—and then repeat the sampling process, thereby collecting a sampling distribution, that distribution will be more or less normal. The bigger the sample, the more normal the distribution will be. This result is true no matter what the shape of the population distribution.

In this demo, you'll get a chance to make any "source" distribution you want, and then make a sampling distribution from that.

What To Do

- Open the file **Central Limit Theorem.ftm**. It will look something like this:

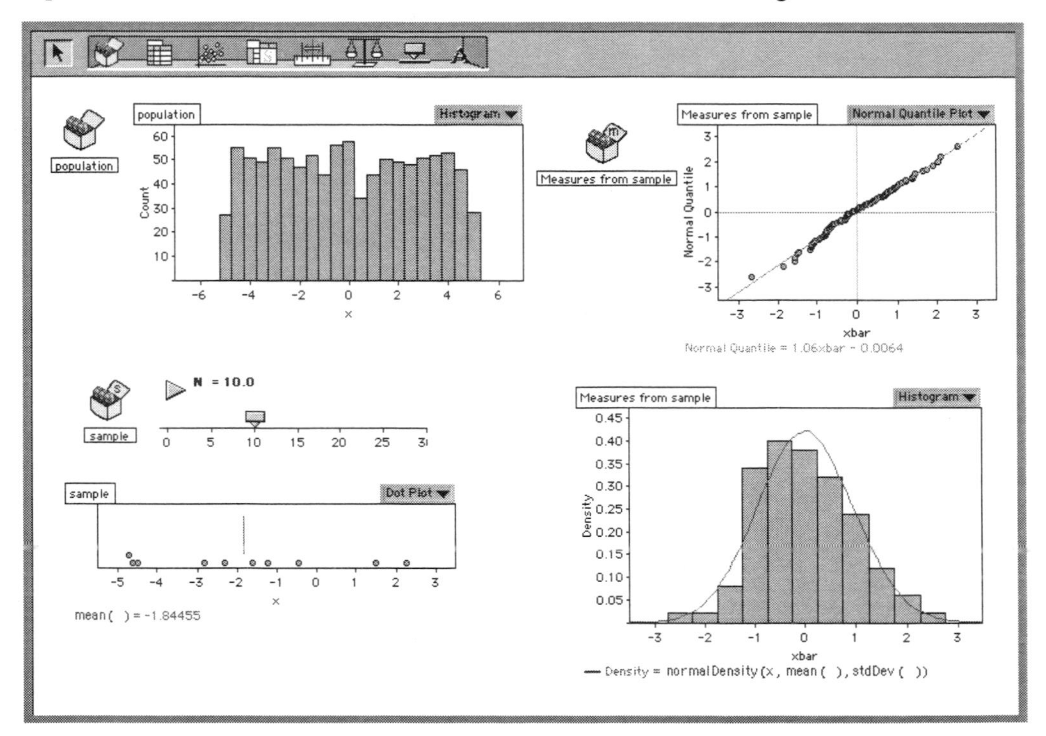

On the left side, you see the **population** above the **sample**. The slider controls how many points appear in the sample. On the right, you see the measures collection—which collects 100 means, called **xbar**, from the sample (i.e., this is a sampling distribution *of the mean*)—and two graphs of **xbar** itself: a normal quantile plot above, and a histogram below, with a normal density function plotted on it.

The shortcut for **Collect More Measures** is **clover-Y** on the Mac or **control-Y** in Windows—but you have to have the measures collection selected for it to work.

- First let's collect a new set of measures. Click once on the measures collection (the gold box called **Measures from sample**) to select it. Then choose **Collect More Measures** from the **Analyze** menu. Fathom collects and displays 100 new values for **xbar**.
- Let's see how different it is for a smaller sample size. Set the slider **N** to 2. Notice that now there are only two points in the **sample** graph. (Note: if you set it to 2.1, say, that's fine; Fathom uses the next lower integer in this case.)
- **Collect More Measures** again. (Select the measures collection, then choose **Collect More Measures** from **Analyze**.) Chances are the graphs will still look approximately normal (or at least triangular, as in "Adding Uniform Random Variables" on page 81).

- So far, our source data has been roughly uniform. Let's change that. Select bars in the top-left **population** histogram, and drag them to the two ends (with the *hand*, as shown; if you use "arrows," you'll just change the bin widths). That is, make a seriously bimodal distribution, as shown in the illustration. We have also rescaled the vertical axis.

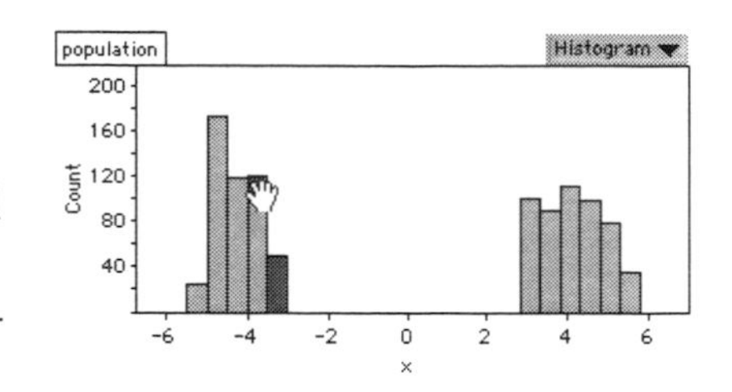

- Now, with **N** still equal to 2, collect measures again and see what you get. You should see three spikes in the **xbar** graph. Not normal at all!

This shows that you don't always get a normal distribution; sometimes you need a larger sample size to make the sampling distribution look more normal.

- So make **N** larger (between 25 and 30 will do) and collect measures again. Shazam! The **xbar** graph looks normal! (More precisely, approaching normal, which is all the Central Limit Theorem guarantees.)
- Even so, we can still make it non-normal. Take most of the points from one of your two humps and move them to the other hump. That is, it should be bimodal, but grossly asymmetrical: lots in one hump, a few in the other. With **N** between 25 and 30, **Collect More Measures** again. You should see something like the illustration—again, not normal.

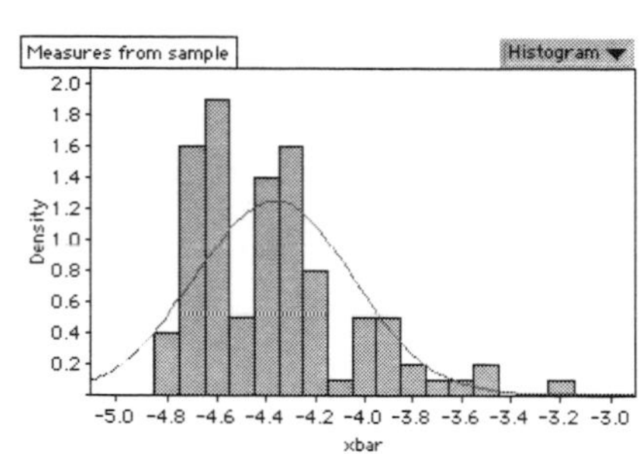

- Sample size to the rescue! Set **N** to about 100, **Collect More Measures** again, and wait patiently for Fathom to do its work. Ahh, probably normality returned (depending on how sharp you made your big spike).

Extensions

Just play with this. Make different distributions and sample sizes, and see when the sampling distributions look normal—and when they don't.

Harder Stuff

1. Select the population collection, then choose **Revert Collection** from the **File** menu to get our uniform distribution back. Now, instead of a sampling distribution of the mean, make a sampling distribution of the *standard deviation* (define it in the **Measures** panel in the inspector for the **population**). Try different sample sizes and population distributions. What do you find?
2. web — Imagine making an asymmetrical, sharply bimodal distribution (just zeros and ones, but lots more zeros). Think about what we found out about what it takes to make the sampling distribution of the mean look normal. How does that relate to the rule of thumb ($np > 10$) for using the normal approximation when you calculate confidence intervals for proportions? (*See* "Why np>10 is a Good Rule of Thumb" on page 107.) In fact, don't just *imagine* it; do it!
3. Imagine (or make) a sharply bimodal distribution. Now consider the sampling distribution for the *median*, and try it out. What do you find?
4. We never changed the *number* of samples in the sampling distribution. What would be different if we collected more points?

Confidence Intervals

One of the main tasks in statistics is to make estimates of parameters. Calling them *estimates* is important: we will never know the parameters exactly, but we can know something *about* them. Confidence intervals (CIs) express what we know. To calculate confidence intervals is straightforward; to know what they really mean, however, is deep and tricky.

CI of a Proportion

While all confidence intervals share the same basic meaning, books often treat the confidence interval for a proportion separately. So do we:

"The Confidence Interval of a Proportion" on page 100. Here we explore one way of defining confidence interval: as the range of population values for which the observed value is *plausible*.

"Capturing with Confidence Intervals" on page 103. In this demo, we see the consequences of the more orthodox definition of the CI: if you construct intervals repeatedly, some of them miss the true value. The fraction that miss depends on the confidence level.

"Where Does That Root(p(1 – p)) Come From?" on page 105. That algebraic snippet is in the orthodox CI formula. Where does it come from? It's the standard deviation of yes/no data mapped onto one and zero.

"Why np>10 is a Good Rule of Thumb" on page 107. They give you these rules, like "use the normal approximation if $np > 10$." This demo shows why that makes sense.

CI of Other Things

Mostly, this is the confidence interval of the *mean*. Some of these parallel their cousins in the proportion section above:

"How the Width of the CI Depends on N" on page 110. This parallels "How the Width of the Sampling Distribution Depends on N" on page 88. We find empirically that it's inversely proportional to the square root of the sample size.

"Using the Bootstrap to Estimate a Parameter" on page 113. If you know nothing about the population distribution, you can still estimate a parameter (a median in this case) using the sample you have. This shows the power of resampling.

"Exploring the Confidence Interval of the Mean" on page 116. If you have only three points, what does the CI look like? We drag the points to find out.

"Capturing the Mean with Confidence Intervals" on page 118. This is parallel to "Capturing with Confidence Intervals" on page 103, but with the mean instead of the proportion. We see how, sometimes, you miss.

Demo 28: The Confidence Interval of a Proportion

Defining the confidence interval • Looking at sample results in terms of plausibility

This demo is all about the meaning of a confidence interval (CI). The CI is a tough concept; it's easy to get muddled. We can go around about how the CI measures confidence in a *process*—the process of constructing the interval—and that the confidence level is the probability that if the process were to be repeated, such-and-such a percentage of the intervals so constructed would contain the true population value. While that's accurate, it sounds circular and is mired in subjunctive, subtle language.

I prefer this definition, which is the basis of our demo:

The confidence interval is the range of possible parameter values for which the observed value is *plausible.*

What do we mean by "plausible?" Let's look at the rest of the statement in context. Suppose we have poll results that say that 56% of people support Measure Q. This poll is of a particular number of people N, the sample size. We assume that it's a simple random sample, and that N is much smaller than the population.

The population parameter we're interested in is p, the proportion of people who favor Measure Q. (In other contexts, we might be interested in the *mean* height of students or the *median* income of dock workers. But this is the confidence interval of a *proportion*; the logic will be the same.) Suppose we polled 50 people and 28 of them said they favored Q. That gives us the sample proportion of 0.56—and is where the 56% came from. We call the sample proportion $\hat{p}$, pronounced *p-hat.* We're interested in the population parameter p, *but we will never know its value.*

Instead, we ask, "Would we be surprised if the true population parameter were 54%?" Certainly not. It would be easy to get 28 positive responses out of 50 if the population had 54% in favor. But would it be plausible for only 10% of the population to be in favor? No way. The chances of getting 28 out of 50 are too small. Traditionally, the border between plausible and implausible is a 5% probability, though you can change that depending on your situation. All that's left is to find that range of values for which the population parameter is plausible.

That's a long introduction; let's see what it means:

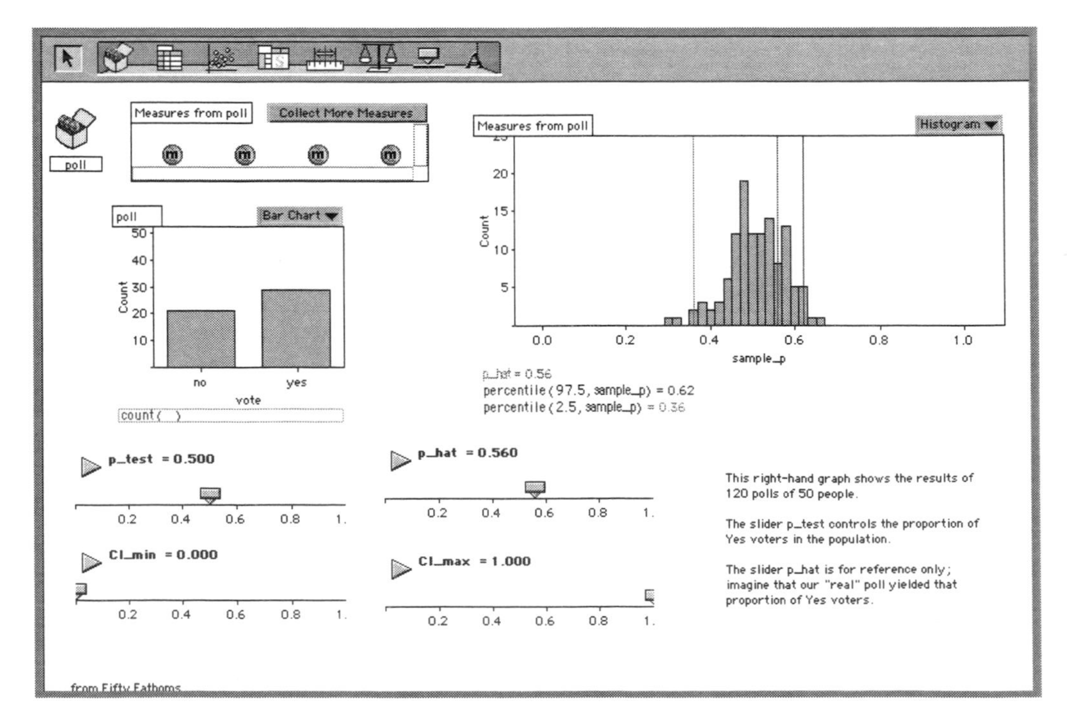

What To Do

- Open the file **CI of a Proportion.ftm**. It will look something like the illustration.

This document simulates many repeated polls. The **poll** collection (upper left) contains 50 cases, **yes** or **no**, randomly controlled by the **p_test** slider. That is, **p_test** is the probability that a respondent in **poll** will vote **yes**. The bar graph shows the results of *one* of those polls.

The **Measures from poll** collection and graph show the results of 120 simulated polls of 50 people. In the graph, the two "outside" vertical lines show the 2.5% and 97.5% percentiles for the polls in the graph; that is, 95% of the polls are between the two lines. We will call all of those results *plausible*.

The **p_hat** slider controls a line on the graph; it represents the results of our real-world poll. So the graph in the illustration, with a population proportion of **p_test = 0.50**, shows that our poll, **p_hat**, is a plausible result.

Note that the histogram of **sample_p**—the simulated poll results—is very much like the graph in "The Distribution of Sample Proportions" on page 80.

Note: if you set **p_test** to unusual values (such as –1) the histogram may look strange. Just choose Histogram again from the menu in the corner of the graph.

- Drag the slider called **p_test**. The graph will *slowly* update (it has to simulate 120 polls of 50 people every time). Set it to 0.3. For that value, you can see that our **p_hat**, 0.56, is outside the bounds of plausibility. If 30% of the true population favored Measure Q, it would be really unlikely that we would get a 56% result in a poll of 50 people.
- Experiment with different values of **p_test** to find values so that the upper and lower "boundary" lines on the graph are coincident with the **p_hat** line. (In fact, since Fathom draws the **p_hat** line first, these "percentile" lines will hide the red **p_hat** line when they match exactly.)

Question

1 Why do you suppose the "percentile" values below the graph seem to be mostly multiples of 2%? That is, you might get 0.56 or 0.58, but seldom 0.57, and never 0.572.

Onward!

- When you find a proportion that seems the minimum **p_test** for which **p_hat** is plausible, set the slider **CI_min** to that value. When you find the maximum value, set **CI_max**.
- Let's draw the theoretical distribution functions for population proportions at those two limits. Select the graph and choose **Plot Function** from the **Graph** menu. The formula editor opens.

For help with this function, see "Plotting Binomial Probability—and Other Discrete Distributions" on page 170.

- Enter **binomialProbability(round(50x), 50,CI_max)*120**. Press **OK** to plot the function and close the editor.
- Choose **Plot Function** again; this time enter **binomialProbability(round(50x), 50,CI_min)*120** (that is, the same except for "min" instead of "max"). Press **OK** to close the editor.
- With the data on the graph, it's pretty busy. Choose **Select All** from the **Edit** menu to select all the cases in the graph; then choose **Delete Cases** from the **Edit** menu to get rid of them.
- We also have those leftover vertical lines. Select the first one's formula at the bottom of the graph (it starts **percentile(97.5,...)** and choose **Clear Formula** from the **Edit** menu. It evaporates. Do the same to the other **percentile** line. Now you have only the functions and **p_hat** left. The graph will look like this:

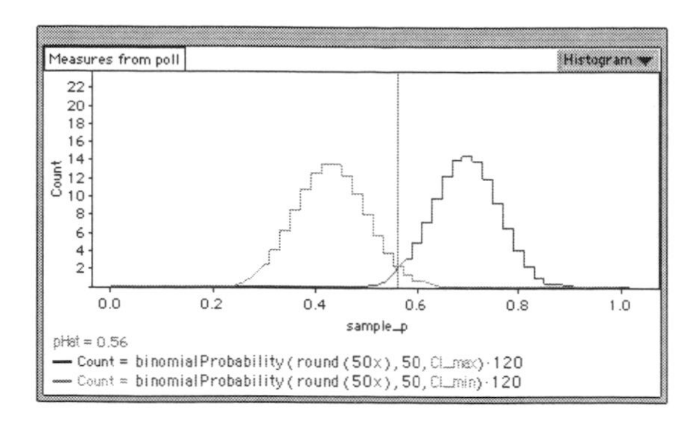

Note: your **p_hat** line may not cut the curves exactly at the intersection.

Now our display shows us how the distributions of the minimum and maximum of the interval relate to **p_hat**. The **p_hat** line cuts off two tails of the distributions; each has an area of approximately 2.5%. Those tails represent the chance that we are wrong, that our interval fails to enclose the true population proportion.

If you want less risk of making that mistake, move the **CI_min** and **CI_max** sliders to make the tail bits smaller. Then the peaks will be farther apart, demonstrating in a different way why confidence intervals get wider—less precise—if you increase the confidence level.

- Move **p_hat** to a different value, then move the **CI_min** and **CI_max** sliders to cut off about the right amount of tails.

Another Question

2 What happens when you set **p_hat** to a very large or small number, like 0.96 or 0.02? What values of **CI_max** and **CI_min** are appropriate?

Extensions

- Test your interval by using an orthodox confidence interval: drag a new estimate off the shelf. (It looks like a ruler.)
- Choose **Estimate Proportion** from the pop-up menu in the estimate itself.
- Enter the data directly into the estimate (do not drag an attribute there). For example, make it look like the illustration.
- Compare your "plausibility interval" values to the ones you see in the estimate.

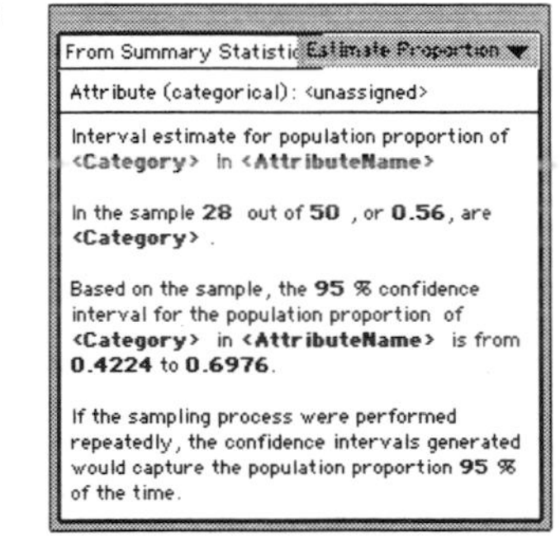

- Change the confidence level to 90% and redo both the plausibility interval (as above) and the one in the estimate.

Harder Stuff

web

1 Explain clearly why this "plausibility interval" is really the same as the orthodox confidence interval ("if you were to draw new samples and construct intervals repeatedly…") described at the beginning of this demo.

2 Explain why, to get a 95% confidence interval, we use the 2.5 and 97.5 percentiles. That makes sense if you have one distribution (the two tails together make 5%), but here we have two distributions with a combined area of 2, not 1.

3 What if **p_hat** is zero or one? Can you still make a "plausibility interval?" Explain what it means, and relate it to the traditional confidence interval and what you get using Fathom's estimate.

Demo 29: Capturing with Confidence Intervals

How confidence intervals of a proportion do not always capture the population value

The mantra of the 90% confidence interval is that if the process were repeated many times, 90% of the intervals we would generate would contain the [unknown] population proportion. (Many teachers are careful to point out that it does *not* mean that there's a 90% *chance* that the interval contains the proportion; other teachers—Bayesian in spirit, perhaps—wonder if there's any difference.)

In this demo, we simulate the situation, generating a lot of intervals, and see how many contain the truth. To do this, we need to know the population proportion, of course. Just remember that in real life, you only get *one* of the possible intervals, and you will have no idea which one it is.

What To Do

- Open the file **Capturing Props with CIs.ftm**. It should look something like this:

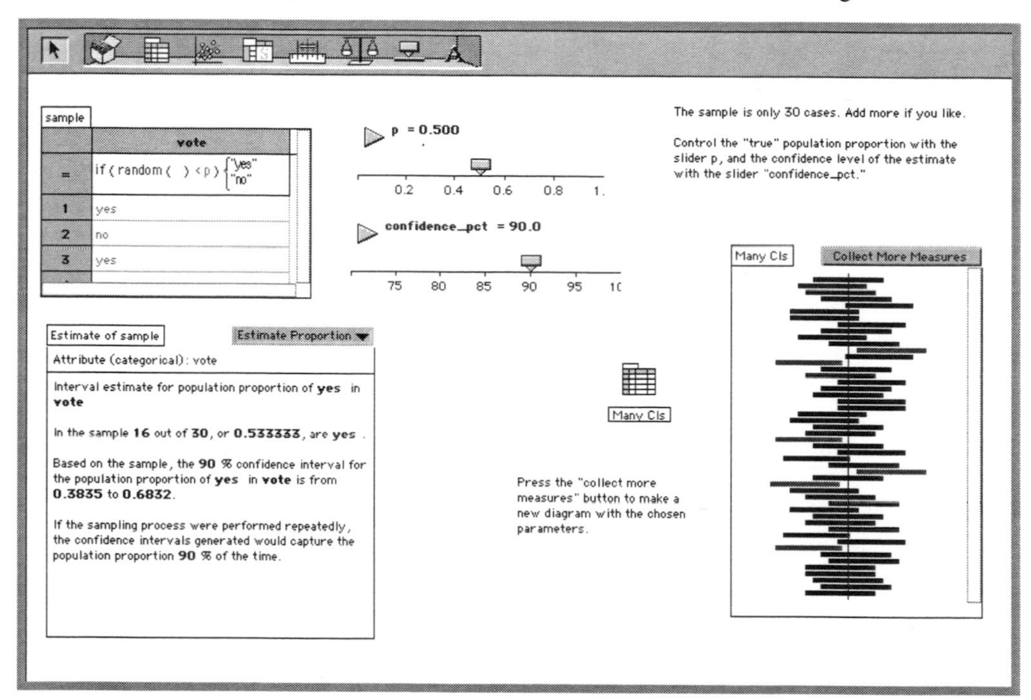

You can see the case table showing the first three of a sample of 30 voters in the upper left. Below it, you can see an *Estimate*, a Fathom object that computes a confidence interval—in this case, a 90% confidence interval. Note the "CI mantra" in the text. At right, we see the results of pulling 50 samples from the same population and computing their confidence intervals. The true proportion—which you would never know in real life—is the vertical line. The red bars are the intervals that missed the true proportion.

Note: **Estimate of sample** (the thing with the CI in the lower left) corresponds to the *last* (bottom) bar drawn in the diagram at the lower right.

- Press the **Collect More Measures** button in the "bar" display (**Many CIs**). The display will recompute, drawing 50 new samples and computing 50 new confidence intervals.
- Repeat the process, noting (informally) how many confidence intervals miss the true proportion.

- Predict what will be different if you change from a 90% confidence interval to a 95% confidence interval.
- To make that change, move the slider named **confidence_pct** to 95 and press **Collect More Measures** again. See if your prediction was correct. Repeat as needed.
- Do the same for an 80% confidence interval: predict, then test, setting **confidence_pct** to 80.
- Predict how it would look different if you had a 90% interval, but with a population proportion of only 0.05.
- Try that out, setting sliders **confidence_pct = 90** and **p = 0.05**.

Questions

1. About how many CIs missed the true proportion when you had the initial settings (**confidence_pct = 90** and **p = 0.5**)?
2. About how much did that number vary from experiment to experiment?
3. When you changed to a 95% confidence interval, what happened?
4. What happened with the 80% intervals?
5. What happened when you used a proportion of **p = 0.05**?

This demo may give you a different look at confidence intervals. First, it's great to see how confidence intervals can miss the true value.

Second, it's great to see how changing the confidence level changes the look of the intervals. When you go from 90% to 80%, the intervals stay in the same places, but they get *smaller*—so the ones that *almost* miss the true proportion start to miss it. This is a confusing issue: it's easy to think, 95% is a better interval—more confidence, after all—so it should be smaller—that is, more precise. Not so.

Finally, when you change the proportion to a value near 0 (or near 1), you can see how the bars are no longer all about the same length. They are also asymmetrical when Fathom uses the exact binomial calculation, but you cannot see that on this display.

We look at how confidence intervals for the mean of a continuous attribute capture (and fail to capture) the true mean in "Capturing the Mean with Confidence Intervals" on page 118.

Extension

Play with the document **CI Stairway 2.ftm**. This shows a famous representation of the confidence interval; follow the hints in the document, whose graph appears at right. Play with the three sliders to see which influence the widths of the bands; figure out how and why.

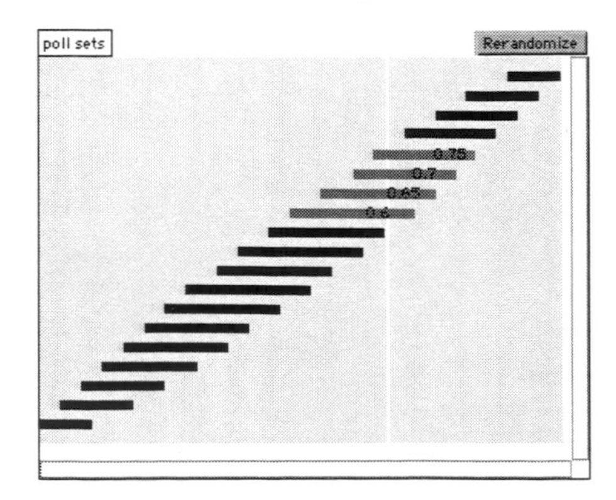

Another document, **CI Stairway.ftm**, does much the same, but it actually generates polls, and so is slower but has sampling variation.

Demo 30: Where Does That Root(p(1 – p)) Come From?

The standard deviation of a variable that's only zero or one • Connecting the "proportion" situation to the "mean" situation

When you study the confidence interval for a proportion, you might learn this formula:

$$CI = \hat{p} \pm z^* \sqrt{\frac{\hat{p}(1-\hat{p})}{n}}$$

Where z^* is the critical z-value (1.96 for a 95% confidence interval), $\hat{p}$ is the sample proportion, and n is the size of the sample. This formula assumes that n is large enough and that $\hat{p}$ is not too close to 0 or 1. (See "Why np>10 is a Good Rule of Thumb" on page 107.)

If you ask the professor what the $\sqrt{p(1-p)}$ means, she may mutter, "It's the standard deviation.[1]" Yet it doesn't *look* like a standard deviation. If you press her, she may confess that "it is the standard deviation if you code a success as one and a failure as zero"—and leave it at that. What does that mean? That's what this demo is all about.

What To Do

- Open the file **SD of a Bernoulli Variable.ftm**. It will look something like this:

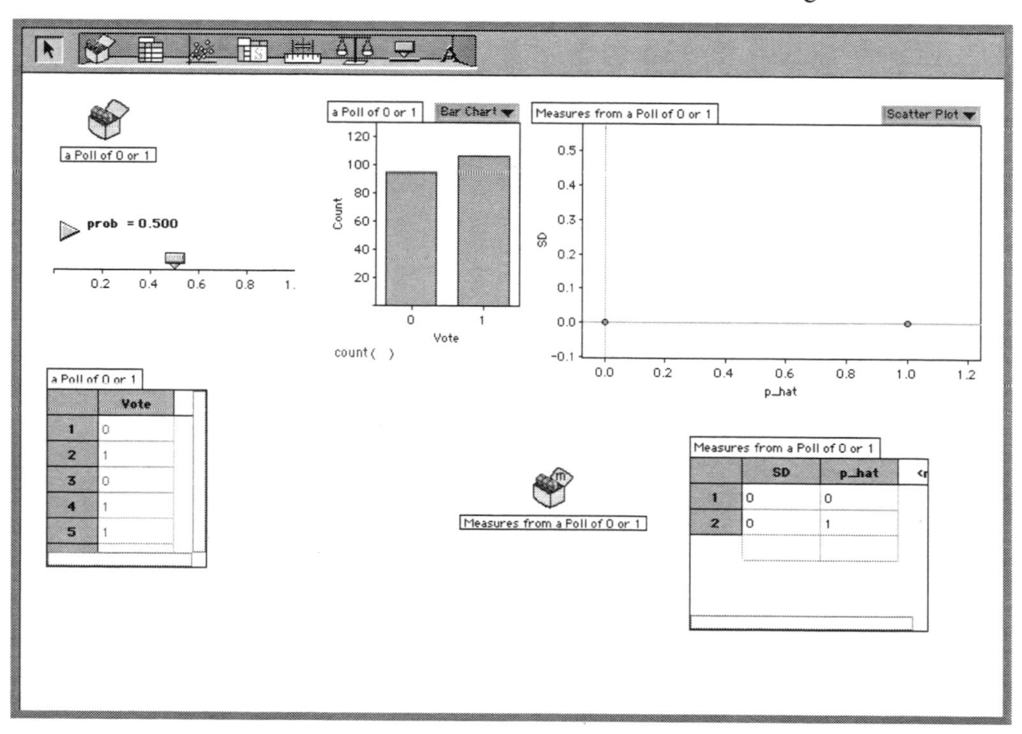

Let's focus on the left half of the document first: it's a simple poll. The collection is 200 cases with one attribute: **Vote**, which is 0 or 1. The slider **prob** determines the chance that a **Vote** will be 1 instead of 0. You can see the results of the poll in the bar chart in the middle. Notice that even though **prob** is 0.500, the counts are not equal.

- Drag the slider and watch how the proportions of **Vote** change in the graph. (Change the bar chart to a ribbon chart if you think that's clearer.)

---

1. Throughout this demo, we ignore thorny issues such as whether we should be using population SD or sample SD, whether we should use $\hat{p}$ or p, etc. In this situation, those distinctions do not matter much. The point is that $\sqrt{p(1-p)}$ measures the spread in the set of 1's and 0's that make up the original data.

- Keep dragging the slider and also notice how the values for **Vote** change in the case table. Note that while you can see only five cases, there are 200 cases in all. (You can stretch the table to see more cases if you like.)

Note: **SD** is **popStdDev(vote)**; since we have 200 cases, this is very close to the sample SD.

Now let's look at the right half. The "measures" collection collects values for **p_hat**—the proportion of 1's—and the standard deviation of the sample, **SD**, whenever we tell it to. You can see that we have already collected two important data points: the standard deviation is zero when **p_hat** is zero or one.

- Now let's tell the measures collection to collect more data. Set the slider **prob** to somewhere near 0.5.
- Click the measures collection once to select it.
- Drag the right edge of the Measures collection to the right just a tiny amount. A **Collect More Measures** button will appear (as shown) as soon as the collection is no longer "iconified."

- Press the button. A new data point appears on the graph (lower right), and the new data appear in the table.
- Repeatedly move the slider and press the button to fill in the arc of points.
- When you're done adding points, put in the curve. First, click on the graph to select it.
- Choose **Plot Function** from the **Graph** menu. The formula editor appears.
- Enter $\sqrt{\text{p\_hat}(1-\text{p\_hat})}$, as shown. Press **OK** to exit the editor. Shazam! The curve appears, and should go right through the points.

Do *not* enter the "SD=" part of the equation—just the part to the right of the equals sign.

So when you have just ones and zeros, you can easily calculate the standard deviation by using that simple formula. The more profound lesson is that the formula for the CI of a proportion is really the same as the one for the mean: *the width of the confidence interval is* $t^* s/\sqrt{N}$. The formulas just look different because of this shortcut for the standard deviation in the proportion case.

Questions

1 When you set **prob** to 0.500, the counts are not equal. Why not?

2 Why is the standard deviation zero when **p_hat** is 0 or 1?

web

3 Does that mean that the confidence interval there has zero width? Why or why not?

Harder Stuff

4 You might wonder why it has to be 0 and 1. Why not 0 and 2? Or −1 and +1? One reason is that then the *mean* of the data is conveniently equal to the proportion. Prove it.

web

5 We demonstrated that $\sqrt{\hat{p}(1-\hat{p})}$ was equal to the (population) standard deviation as computed by the computer. Let's check that analytically. You know that the standard deviation is really the square root of the variance, and that variance is the expected value of the squared deviation from the mean, that is,

$$Var[X] = E[(X-\mu)^2].$$

Show that this is equal to $p(1-p)$.

Demo 31: Why np>10 is a Good Rule of Thumb

Explaining the np > 10 rule for using the normal approximation in the CI of a proportion

When you find a confidence interval for a proportion, there is a question of whether the normal approximation is correct. What does that mean? For one thing, the proportions are like a population of ones and zeros. So the distribution of sample proportions will be binomial. The orthodox confidence interval formula,

$$CI = \hat{p} \pm z^* \sqrt{\frac{\hat{p}(1-\hat{p})}{n}},$$

is based on the idea that the sample proportions will be normally distributed. But didn't we just say that they were binomially distributed? Sure. But the binomial distribution looks pretty normal as long as n is large enough and $\hat{p}$ is far enough from 0 or 1.

But how far is far enough? The rule of thumb is that $np > 10$ [1]. But where does that come from? One way to study this question is to ask: What happens if np is small? What can happen if you keep using the normal approximation? For one thing, you can calculate confidence intervals for proportions that extend *below zero*. And that's bad. After all, a confidence interval is the range of proportions that could plausibly give rise to our sample. Are we willing to say that the proportion of "yes" voters might easily be –0.08? I don't think so.

In this demo, you control values for n and $\hat{p}$, and Fathom generates lower bounds for confidence intervals using the normal approximation, plotting them on a separate graph. They're coded to show whether the interval extends below zero or not.

We're looking to figure out what values for n and $\hat{p}$ give "good" confidence intervals. Let's begin!

What To Do

- Open the file **Why np is greater than 10.ftm**. It will look like this:

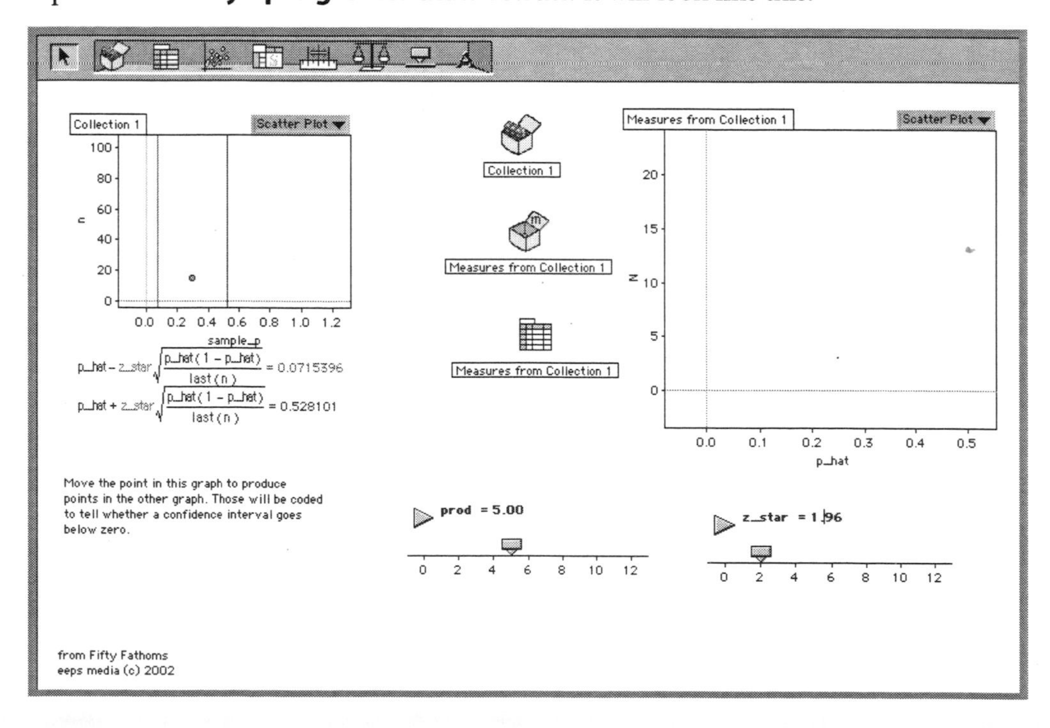

1. Also $n(1-p) > 10$, but we'll just focus on the bottom end.

- Grab the single point in the **Collection 1** graph at the left. This point is your controller; by dragging, you change the values for n and p.[1] The confidence interval extends from the red line to the blue line. Note: the axis bounds for this graph are pretty good; you shouldn't need to change them. Those of the other graph may jump around. Don't let that bother you.
- Carefully move the point down and to the left (towards the origin). Points will appear automatically in the larger graph on the right, mirroring your motion in the small graph.

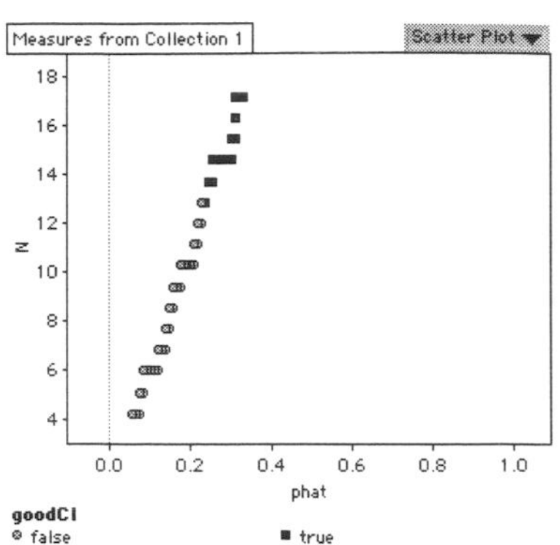

- Drag until you can see two different symbols in the big graph, and a legend appearing for the attribute **GoodCI**, which is either **true** or **false**. In the illustration, the blue squares are good, and the gray circles represent bad confidence intervals—where the interval extends below $p = 0$.
- Continue to drag the point around, in different directions, until you get a pretty good idea of what part of the plane is "good" and what part is not. You should see a region near the axes that's bad. That is, if n is small or p is small, it may mean that the normal approximation stinks.
- Let's write a function that approximates the border of this region. Click once on the larger graph—the one with lots of points—to select it.

- Choose **Plot Function** from the **Graph** menu.
- In the formula editor, enter "**prod/p_hat**," as shown. Then press **OK** to exit the editor. A curve appears—a rectangular hyperbola.
- Drag the **prod** slider to change the curve. Make it so that the curve roughly matches the boundary between good confidence intervals and bad ones. You should get a value something like 3.

This value, three, is *the smallest product of the sample size and the sample proportion* you can possibly imagine using with the normal approximation. If $np < 3$, you're guaranteed a crummy CI—one that extends below zero.

Questions

1. Suppose you did a poll of 50 people and 5 of them said their favorite ice cream was pistachio. Where is that point on the graphs? Is it in the "good" area? What's the (normal approximation) confidence interval for that proportion?
2. We said that $np < 3$ was terrible. And we got that by comparing the points to a function, **N = prod/p_hat**. What's the connection between these two? That is, why is one multiplication and the other division?

Extension

But isn't the rule of thumb $np > 10$? Sure. We have just found the bare minimum here. We know that $np < 3$ is bad; but that doesn't mean $np > 3$ is good. For one thing, we have been identifying "good" as any CI that does not include zero. This is a little strange. Suppose I take a sample of 20 and get four successes. The lower bound of the 95% confidence interval (with the normal approximation) is about 0.02. Is that plausible?

---

1. Basically, we have created a two-dimensional slider out of a point. See "Using a Point As a Controller" on page 168 for how we did it.

Hardly. If I had a process with a 2% chance of success, and I did it 20 times, I'd get 4 or more successes fewer than one time in a thousand.

So what shall we do to improve our rule of thumb? One way is to remember that we've only been looking at 95% confidence intervals—ones with $z^* = 1.96$. Control this value with the slider named **z_star**. Bump that up to about 3—that is, look at a CI width of 3 standard errors instead of 2—and see what happens.

Another Question

web

3 If you take the sample size and multiply it by the sample proportion, don't you get the number of successes? That is, could you say that the rule of thumb "*np*>10" really means that you have to get at least 10 people to say yes (and at least 10 people to say no) before you can use the normal approximation to the CI?

Another Extension

We have really just looked at whether the normal-approximation confidence interval overlaps zero. But another question is whether the overall shape of the distribution matches the binomial. Open the file **Binomial v Normal.ftm** and explore the difference. It looks like this:

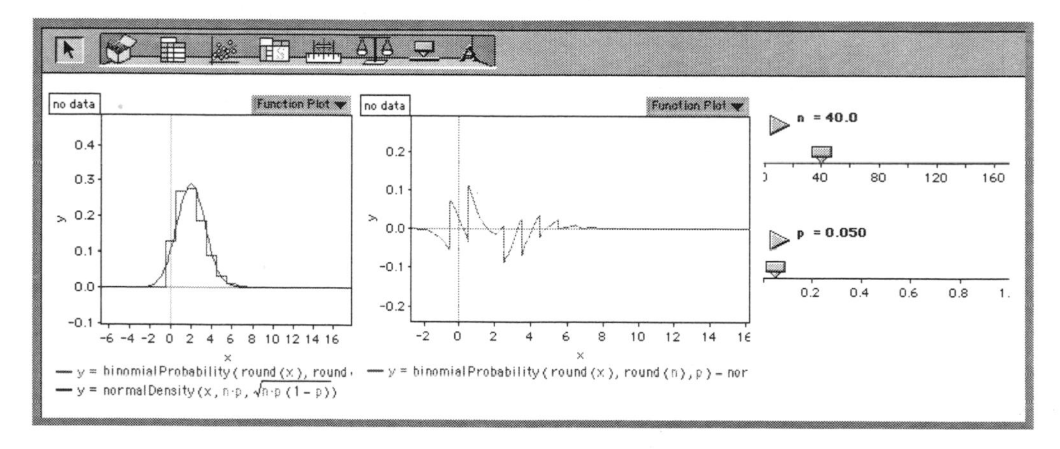

You can see, at the left, the normal and binomial density functions plotted together. At right, you see the difference between those functions. The sliders **n** and **p**, control the sample size and the probability of success, respectively. If you make **n*p** large, the curves match well, and the difference is small. If **n*p** is small, the difference goes nuts, *whether or not much of the normal function lies below zero.*

Harder Stuff

4 Show that the assertion above— "If I had a process with a 2% chance of success, and I did it 20 times, I'd get 4 or more successes fewer than one time in a thousand."—is true.

5 If you have a computer to help you, you don't need the normal approximation—the binomial distribution gives the correct answer. Explain why the binomial is always correct.

web

6 Perform an experiment to test how much difference it makes to use the normal approximation instead of the binomial. Use a sample size of 10 and a true population proportion of 0.2. Draw repeated samples and construct 95% confidence intervals using both the binomial estimate and the normal approximation. See how many CIs capture the true proportion, and on which side they miss. Vary the sample size and the population proportion.

Demo 32: How the Width of the CI Depends on N

How the width of a confidence interval is inversely proportional to the square root of the sample size

In this demo, instead of making the distributions of means as we did in “How the Width of the Sampling Distribution Depends on N” on page 88, we will start with a Fathom estimate, that is, we’ll compute a confidence interval based on our sample. Then we’ll track the width of the interval as the sample increases in size.

What To Do

Open the file **Width of CI depends on N.ftm**. It should look something like this:

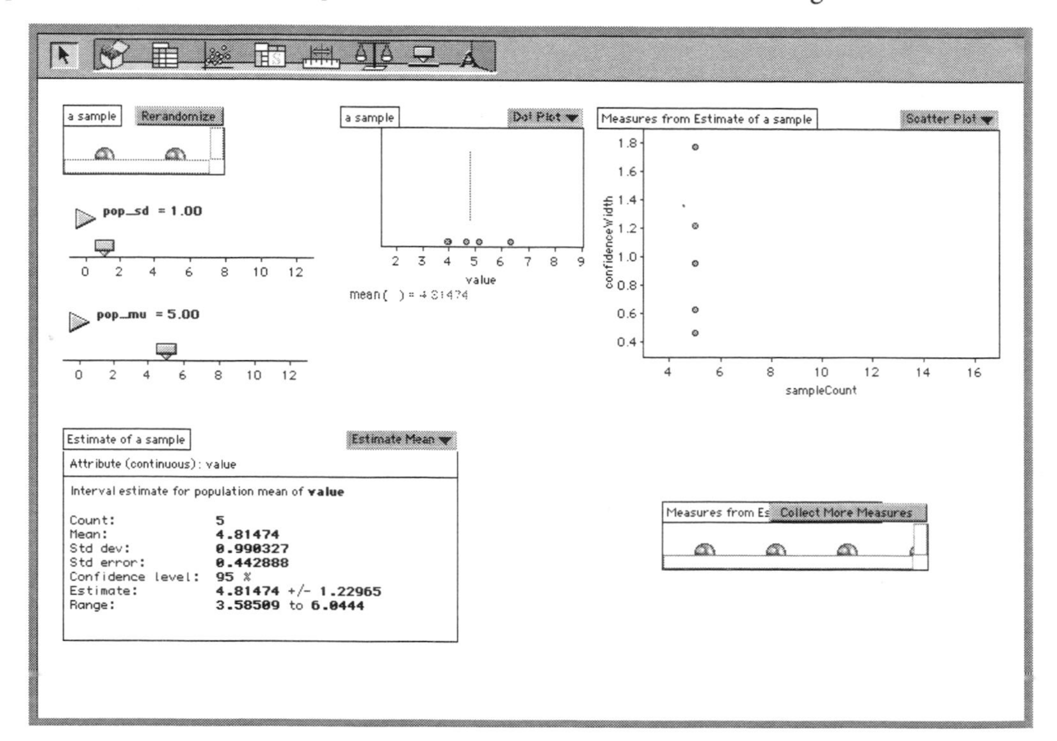

The collection **a sample** currently has five cases, drawn from a normal population with mean **pop_mu** and standard deviation **pop_sd**. They’re graphed in the dot plot top center. At the lower left, we have an Estimate—a Fathom object that, in this case, computes a 95% confidence interval for the mean.

On the right-hand side, the measures collection and the graph above it collect and display information about repeated tests on new samples. So you can see that for a **sampleCount** of 5, there have been five tests. The width of their confidence intervals has ranged from about 0.4 to about 1.8. Note: **confidenceWidth** is the *half*-width of the entire interval.

- Press **Rerandomize** in the **a sample** collection a few times to see how the sample changes. The test changes too; see how the width of the confidence interval jumps around.
- Let’s increase our sample size from 5 to 10. With the collection **a sample** selected, choose **New Cases...** from the **Data** menu. Add 5 more for a total of 10. You’ll see the middle dot plot update.
- Now lets record some confidence intervals. Press the **Collect More Measures** button in the measures collection lower right. Fathom makes five CIs from samples of 10 and adds their **confidenceWidth**s to the graph upper right.

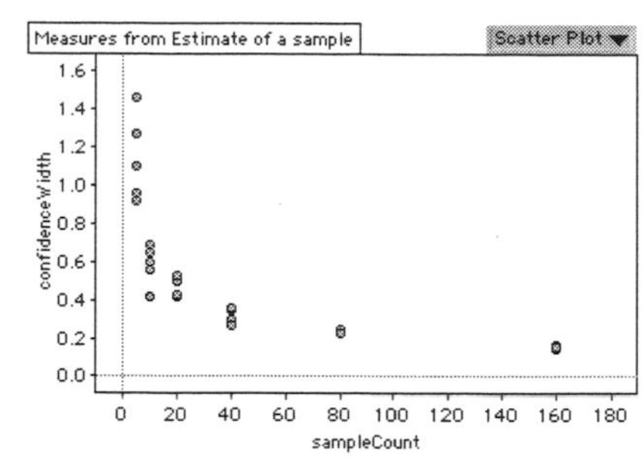

- Alternate between adding cases and collecting measures. Add 10 more cases for a total of 20, 20 more for a total of 40, 40 more for 80, and 80 more for a total of 160. You don't have to do exactly these numbers. Be sure to **Collect More Measures** every time you add cases to the **a sample** collection. When you're done, your graph will look something like the one in the illustration.

Note: at this point in the demo, depending on experience, it's great if you can try different functions to fit these points. An inverse dependence or a decreasing exponential are popular candidates. For brevity, we omit the specific instructions. One can also use logarithms; see the extensions.

- Let's make a function and see if the one-over-root-N dependence is at work here. Select the graph by clicking on it once. Then choose **Plot Function** from the **Graph** menu. The formula editor appears.
- Enter $1/(\sqrt{sampleCount})$ and press **OK**. The graph will miss, but it's about the right shape.

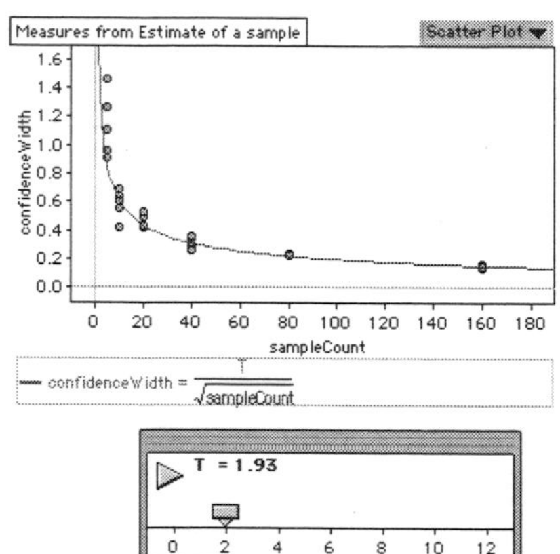

- Let's make a variable parameter so we can fix the function. Drag a slider off the shelf and put it in the blank area below the graph. Edit its name **V1** to be **T**.
- Double-click the formula for **confidenceWidth** at the bottom of the graph to open the formula editor again.
- Put **T** in the numerator. Again, press **OK** to close the editor.
- Use the slider to change **T** so the curve fits the data. You may want to zoom into the cluster of points over 160 to make sure you hit them. Your graph and slider will look something like the ones in the illustration.

It is good to stop here and reflect on what an inverse-root function means: since 160 is four times 40, the width of the confidence interval will be $1/(\sqrt{4})$ or one-half the size. Similarly, by increasing the sample size from 5 to 80, you reduce the confidence width by a factor of 4.

Questions

web

1. How large a sample would we need to reduce the "margin of error" to 0.01? That is, how many points do we need in order to get **confidenceWidth** as low as 0.01?
2. What would be different if we changed **pop_mu**?
3. What would be different if we changed **pop_sd**?
4. What would be different of we changed the confidence level in the test from 95% to, say, 80%?
5. What would be different if we did 10 tests at every sample size instead of just 5?

web

6. What does it mean that the values of **confidenceWidth** get closer together as **sampleCount** increases?

Extension

If you didn't know about the root-n stuff, you could use logarithms to discover it.

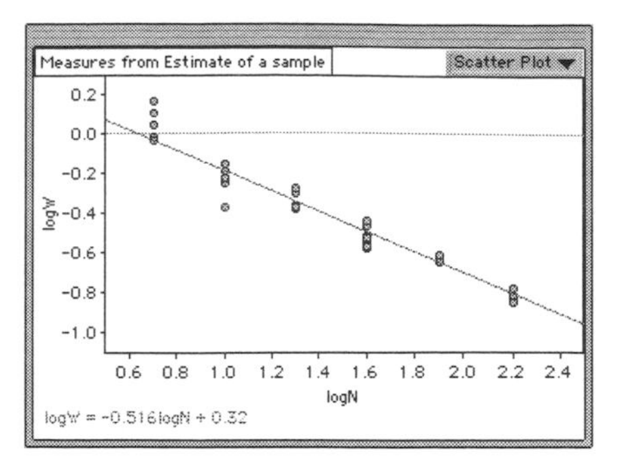

- Double-click the measures collection to open its inspector. Click the **Cases** tab to go to the **Cases** panel.
- Make two new attributes, **logN** and **logW**.
- Give them formulas[1] so that they are the logarithm of **sampleCount** and the logarithm of **confidenceWidth**, respectively.
- Then plot **logN** against **logW** and fit a movable line.[2] (Ignore the points with a sample size of 5; see the harder stuff, below.) You'll get a graph like the one in the illustration. Note the slope; figure out what it means.

Harder Stuff

1 The questions above (under "Questions") each imply an activity; try them. For example, redo this demo with **pop_sd** equal to 2 instead of 1. What happens?

2 It may not have been clear in the original, curvy graph, but it's really obvious in the log-log graph above that the tests with **sampleCount = 5** give a larger **confidenceWidth** than you would expect based on the pattern from the other sample sizes. Explain why.

Tautology Alert

The astute and experienced reader may cry "foul!" after reading and exploring this demo. We discovered that if you make confidence intervals for different sample sizes (all drawn from the same population), the width of the CI is pretty closely inversely proportional to the square root of the sample size. In fact, the fit is so good, you may be suspicious.

Your suspicions will be confirmed when you realize (aha!) that Fathom computes the width of the CI to be

$$w = t^* \frac{s}{\sqrt{n}}$$

where t^* is the relevant t-value and s is the sample standard deviation. So of course it will have a one-over-root-n dependence. So why include this demo? Several reasons:

Most importantly, to approach the issue using a different technique than we did earlier in "How the Width of the Sampling Distribution Depends on N" on page 88. Sure, it's the same principle, but we do not always recognize that the same principle applies when we dress it up in different clothes.

Secondly, knowing that there's a root n in the denominator of the function doesn't mean we have a feel for what that really means. So actually making samples—with real variation—and testing them repeatedly makes good sense.

Finally, and most subtly, since t^* depends on n, the width is *not* in fact inversely proportional to the square root of the sample size; the departure from that rule is exactly why we need t instead of just z.

---

1. e.g., **log(sampleCount)**
2. To get a movable line, first select the graph, then choose **Movable Line** from the **Graph** menu. Drag parts of the line to change its slope and intercept; the equation appears at the bottom of the graph.

Demo 33: Using the Bootstrap to Estimate a Parameter

The bootstrap • Using resampling (with replacement) to create an interval for a parameter

The traditional way to estimate a mean—take a sample from the population and make a confidence interval—assumes that your sample is either large or normal. The Central Limit Theorem makes this process more or less correct for many distributions of data and for some parameters other than the mean, but sometimes (as we might see in "The Central Limit Theorem" on page 96) the distribution is so odd as to make things problematic. Or perhaps you don't want to estimate the mean, but rather some other statistic, and you're not sure that the Central Limit Theorem applies. Or maybe you're just skeptical whether your situation is appropriate for the traditional CI and are looking for something that doesn't make those assumptions. If any of these is true, the bootstrap may be for you.

The key idea behind bootstrapping is this: when you take a sample, that's all the data you have. So you assume that *the distribution of the values in the population is identical to the distribution in the sample*. That's the key point. You don't assume anything about normality.

And then, to find out how far off you might be, you resample from that distribution, calculating the parameter you're interested in. Let's try it, and estimate the *median* income of a population. We have a sample of incomes from the 1990 Census in Champaign County, Illinois.

What To Do

Open the file **Bootstrap.ftm**. It should look like this:

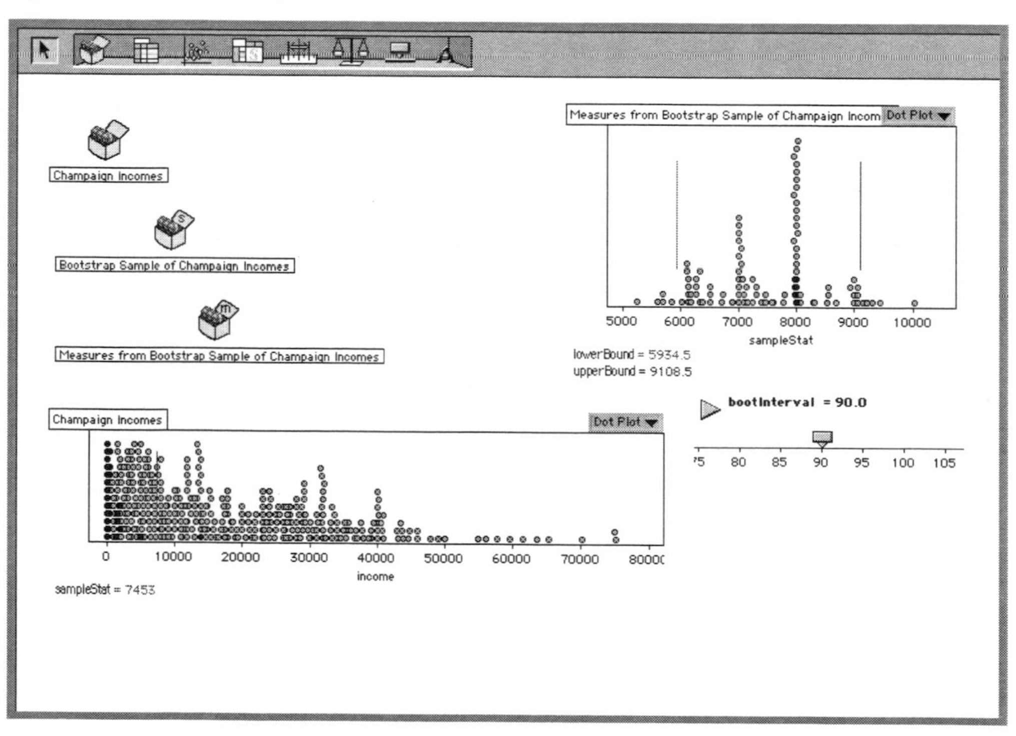

The top collection is the sample of 500 incomes. You can see them in a graph at the bottom. The median (called **sampleStat**) shows up as a vertical line on the graph. The middle collection, **Bootstrap Sample of Champaign Incomes**, is a collection (we did not make a graph of these incomes, but you could plot them if you wanted) where we have sampled from the former collection 500 times with replacement. That is, it's a lot like the original, but some cases will be duplicated and some omitted.

Finally, the bottom collection contains measures from that sample, where, in this case, the measure (**sampleStat**) is the sample *median*. That statistic is plotted in the upper right—a sampling distribution, for 100 repeated bootstrap samples. We have also plotted two values: an upper and lower bound of what's called a bootstrap interval. They are, in this case, the 5th and 95th percentile of that distribution. The percentiles are controlled, in turn, by the slider **bootInterval**, currently set to 90. That is, the two lines encompass the middle 90% of the distribution. So the bootstrap interval is kind of like a confidence interval.

- Let's collect another set of bootstrap measures. Click once on the measures collection (the bottom one) to select it. Then choose **Collect More Measures** from the **Analyze** menu. Fathom gradually collects 100 measures and replaces the data on the graph with the new sample medians.
- Repeat as necessary. Become convinced that, while the upper and lower bounds move around a little, they stay roughly in the same place.
- Drag the **bootInterval** slider (or edit the number) to see what happens.

Questions

web

1. If the median of the original sample is 7453, how can it be that so many of the bootstrap samples have a median of 8000? Shouldn't 7453 be, if not the most popular value, at least close?
2. How can you predict, before you move the slider, which way the bounds will move?

Onward!

Let's make a bootstrap of something else. How about the 75th percentile of income?

- Double-click the source collection—**Champaign Incomes**—to open its inspector.
- Click on the **Measures** tab to bring that panel to the front.
- Double-click the formula for **sampleStat** (it's currently **median(income)**) to open the formula editor.
- Edit the formula to read **percentile(75, income)**, as shown. Press **OK** to close the editor. Note how the line moved in the bottom graph to indicate the new value of the sample statistic.

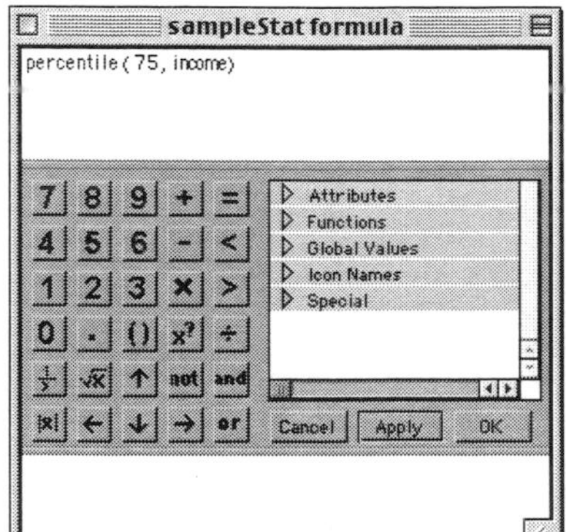

- Collect bootstrap measures again as before. Note the width and position of the 90% interval. (Remember, you control the interval with the **bootInterval** slider.)
- Try any other statistic that interests you. If you can't think of any, here's one: **median(income, sex = "M") – median(income, sex = "F")**.

Harder Stuff

web

3. Explain why the median income of Champaign County seems to be so low. Can it be that half the people earned less than $7500 per year, even way back in 1989?
4. Related task: Look up the median income of Champaign County, Illinois from the 1990 Census. (Use the Internet. Start at **http://www.census.gov**.) Compare it to your interval and explain why it's so far off.

5 Make bootstrap estimates of men and women separately. Do the 90% bootstrap intervals of the median overlap?

6 Make a bootstrap estimate of the *mean* income at the 90% level. Compare that to an orthodox confidence interval (Use a Fathom estimate; choose **Estimate Mean** from the popup menu in the estimate, and drag the **income** attribute from the source collection to the place indicated. Also be sure to make it a 90% interval instead of the default 95%.) How close are they?

7 Explain why a bootstrap interval is pretty much the same as a confidence interval. Refer to the definition of a confidence interval if you have trouble.

8 Describe some advantages and disadvantages of using this bootstrap technique for making estimates of parameters.

Demo 34: Exploring the Confidence Interval of the Mean

How the CI depends on individual values

This demo gives you a chance to see (or show) dynamically how the confidence interval changes as you change the data values, the confidence level, and the number of points. Its purpose is more to help you get a "feel" for confidence intervals than to reveal their deeper meaning (as in, e.g., "Capturing the Mean with Confidence Intervals" on page 118).

What To Do

- Open the file **CI of the Mean.ftm**. It will look like this:

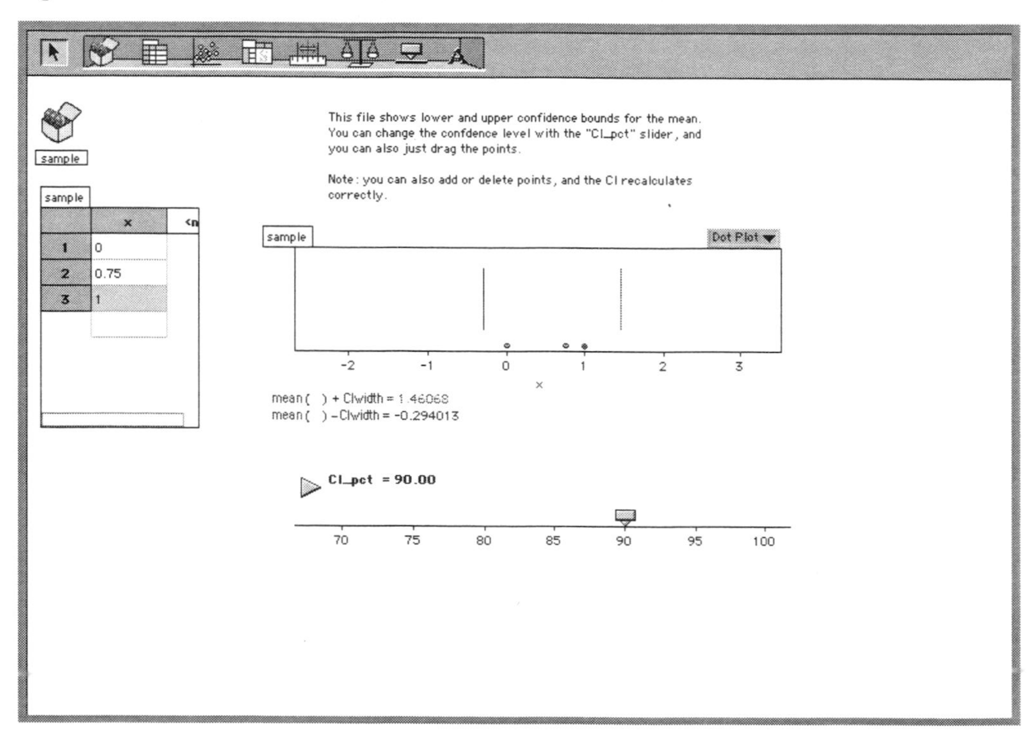

- Pick a point on the graph and drag it. You'll see the confidence interval update in real time. Rescale the axes when you need to. Pay attention to how moving a point changes the width of the interval.
- Drag a dashed rectangle around all of the points to select all three. Then drag one to move them all in parallel. See how the width of the interval does not change.
- See if you can arrange the points so that the interval does not include zero.
- See if you can arrange the points so that one of them is outside the interval.
- The **CI_pct** slider controls the "confidence level" of the interval, in percent. It begins as a 90% confidence interval; change it to the traditional 95%. Then just drag the slider value around and see what happens.

Questions

1. When you move all three points in parallel, what happens to the interval?
2. When you move just one point, what happens to the interval?
3. What seems to determine the width of the interval?
4. Where are the points when the interval does not include zero? (web)
5. When you make the confidence level higher with the slider, what happens to the interval?

Extension

We've been looking at a sample size of three; let's add a point.

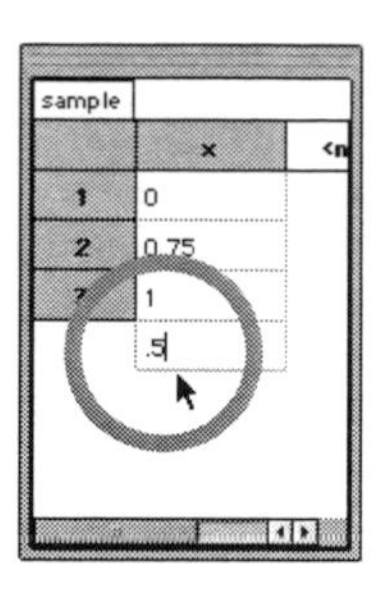

- In the case table at the left of the window, type a new value, as shown, in the empty box at the bottom and press **enter** or **return**. The number should be about the same size as the current data values.
- Redo the steps above with four cases instead of three.
- Add additional cases as you wish.
- To remove a case, select it by clicking on it once in the graph or in the case table, then choose **Delete Case** from the **Edit** menu.

Harder Stuff

6. Find two arrangements of points, as different as possible, that have the same confidence interval.
7. Getting a point outside the CI is hard with three points. Figure out what has to be true for that to occur.
8. Getting a point outside the CI is relatively easy with four points. Figure out what has to be true to make that hard (or even impossible).

Demo 35: Capturing the Mean with Confidence Intervals

How confidence intervals of a mean do not always capture the population value • What repeated CIs look like

In "Capturing with Confidence Intervals" on page 103, we generated confidence intervals for a proportion. We saw that, for a 90% interval, roughly 90% of the intervals encompassed the true value. Here, we do the same, but with confidence intervals for the mean. Again, remember that in real life, we don't know the true mean—and we never will. In practice you get *one* sample and its interval, and you never see the others.

What To Do

Open the file **Capturing the Mean with CIs.ftm**. It should look something like this:

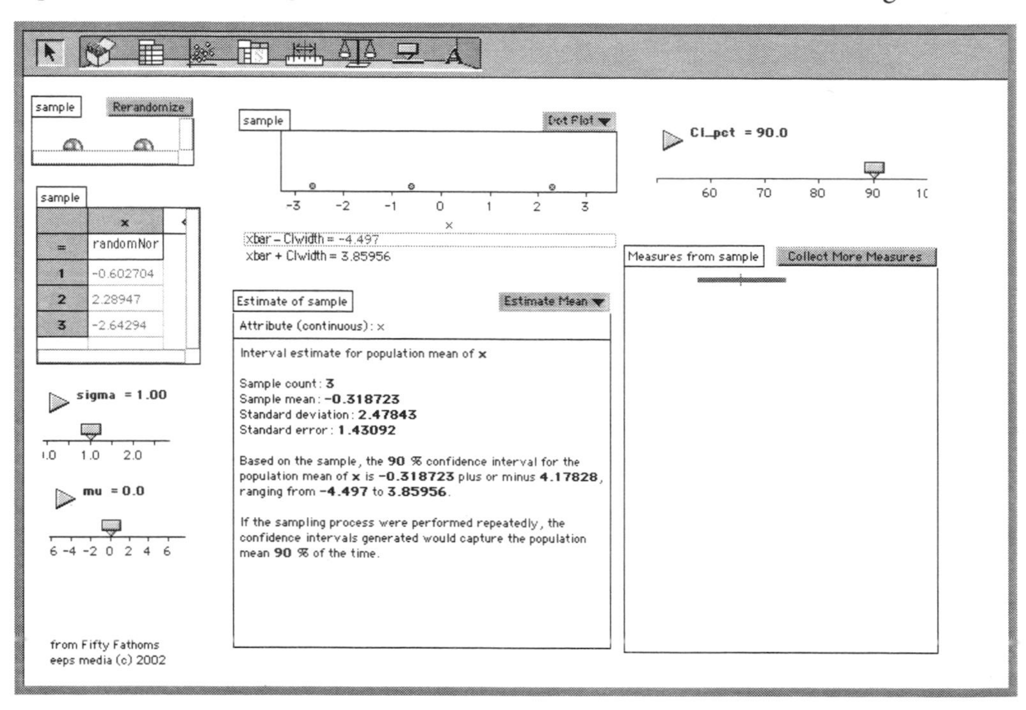

The **sample** collection in the upper left has three cases; we're choosing a sample of three from a normal population with a mean of **mu** and a standard deviation of **sigma**, both controlled by the sliders below the table. The data appear in the graph top center, with the confidence interval limits in blue and red. The **CI_pct** slider controls the confidence level for the interval.

Meanwhile, down below, you see a Fathom estimate ("**Estimate of sample**"), where the program calculates the confidence interval (CI) and gives you additional statistics as well as the text "mantra" of what the CI means. In the lower right is another view ("**Measures from sample**"), this time showing the CI from the estimate as a graphic box; the vertical line is the axis. As they appear, the boxes will be green if they enclose the true mean, black if they don't. We'll accumulate more throughout this demo.

Note: Unlike in "Exploring the Confidence Interval of the Mean" on page 116, you can't drag the points in the graph, because they're calculated in a formula. In fact, that's one reason why we included both of these demos—it's great to drag the points, but it's also great to see what happens with repeated sampling.

Click the **Rerandomize** button in the upper-left collection to see the sample points—and their CI—change. And each time you click, the new interval appears in the collection at the lower right. Explore this until you're comfortable with what's going on.

Note: The "bars" display is really the "gold balls" view of the measures collection; the balls have just mutated into the more useful bars. If you want, check out the **Display** panel of that collection's inspector.

- Now we'll set up Fathom to collect these automatically. Double-click the lower-right display (the green and black bars) to open that collection's inspector. It should open to the **Collect Measures** panel. Edit the panel so that Fathom collects 51 measures (not 1); so that it empties the collection; and so that it does not collect measures when the source changes. The inspector should end up looking like the one in the illustration.
- Close the inspector and test it by pressing the **Collect More Measures** button. It should (gradually) fill with 50 bars (not 51—the first case is the "axis") as Fathom resamples, updating the graph and the estimate 50 times.
- Repeat that until you're convinced that about 90% of the bars encompass the mean.
- Play with the **CI_pct** slider to see what that does to the interval in the various graphs and to the number of intervals that catch the mean.

Questions

1 You would expect that 5 intervals out of every 50 would "miss" the mean when **CI_pct** is set to 90%. But that number should vary. Roughly, what's the range of numbers of "missing" intervals you found?

web

2 In general, what characterizes the intervals that miss (besides the fact that they miss)? That is, what do most of them have in common?

3 When you change **CI_pct** and watch the display of the intervals in the lower right, what happens to the *positions* of the bars? What happens to their *widths*?

web

4 As you drag **CI_pct** to lower values, what happens to the number of intervals that miss? Explain why that is, based on watching the display of the 50 intervals?

Extension

- Change **mu** and **sigma**. What effect does that have on how this works?
- Change the sample size by adding cases to the **sample** collection. (Select the sample collection and choose **New Cases** from the **Data** menu.) What effect does that have?

Harder Stuff

web

5 When we looked at intervals of proportions in "Capturing with Confidence Intervals" on page 103, the intervals were all roughly the same length. Here they differed. Why? Here, the short ones tended to miss. There, we got the same proportion of misses even though none were so short. How could that be?

6 Consider the first question—about the range of numbers of intervals that miss. Record those numbers as data in Fathom, and figure out a confidence interval for that number. See if it encloses the expected "true" value (5 for a 90% interval). Should you use a confidence interval of the mean, or of a proportion, or something else?

Hypothesis Tests

Hypothesis tests and estimates (i.e., confidence intervals) together constitute the backbone of inferential statistics. In Fathom, we can show the meaning of the hypothesis test through simulation. First we simulate the null hypothesis. Then we create a sampling distribution of a statistic where the null hypothesis is true. Finally, we compare our test statistic to that distribution.

Alternatively, we can use canned tests—such as the *t* test of mean—and see dynamically how changing the data changes the results of the test. We can also test repeatedly, using the results as data so that we can study the behavior of the test as a whole.

This section includes:

"Fair and Unfair Dice" on page 122. How can you tell if a die is unfair? We get a simulated, loaded die. First we devise a statistic of unfairness. Then we create the sampling distribution of that statistic for *fair* dice. Comparing the two, we can assess how likely it is that a fair die would behave like our loaded one.

"Scrambling to Compare Means" on page 126. Here we have fake plant data, using two different fertilizers. Is the new one better? We devise a statistic, use scrambling (a randomization technique) to create a sampling distribution, and compare. This is another example of the basic meaning of a hypothesis test.

"Using a t-Test to Compare Means" on page 129. We explored the *t* statistic in the section "Standard Deviation, Standard Error, and Student's t" on page 65. Here we use the test instead of scrambling to compare the same two groups we compared in the previous demo. We will also see how changing the data changes the results.

"Another Look at a t-Test" on page 131. Here we look at repeated *t*-test results as data, and see how sample mean and standard deviation relate to the *P*-value of the test. We also see how random variation produces Type I errors.

"On the Equivalence of Tests and Estimates" on page 133. This quick demo shows how these hypothesis tests give some of the same information as confidence intervals. If the significance levels are equivalent, rejecting the hypothesis is the same as being outside the CI.

"Paired Versus Unpaired" on page 135. When you have repeated measures, is it better to do a paired test or an unpaired test? Paired is better, and this demo shows why.

"Analysis of Variance" on page 137. We return to the basic meaning of the hypothesis test for this one, first devising a statistic to tell—for more than two groups—if a measurement is independent of group membership, then creating the sampling distribution in the case of the null hypothesis for comparison.

Demo 36: Fair and Unfair Dice

Creating a measure of "fairness" • Sampling distributions • Testing hypotheses empirically • The chi-square statistic

Suppose you have a test die, and you roll it many times, and you get suspicious. You think it might be unfair. How do you check? Compare it to dice that you know are fair. But even fair dice give uneven distributions, as you will see. The key is to figure out how to measure the unevenness of a distribution, and then to compare the unevenness of your test die's distribution to a sampling distribution of the same statistic for the fair dice. Sound confusing?

What To Do

- Open the file **Fair and Unfair Dice.ftm**. It will look something like this:

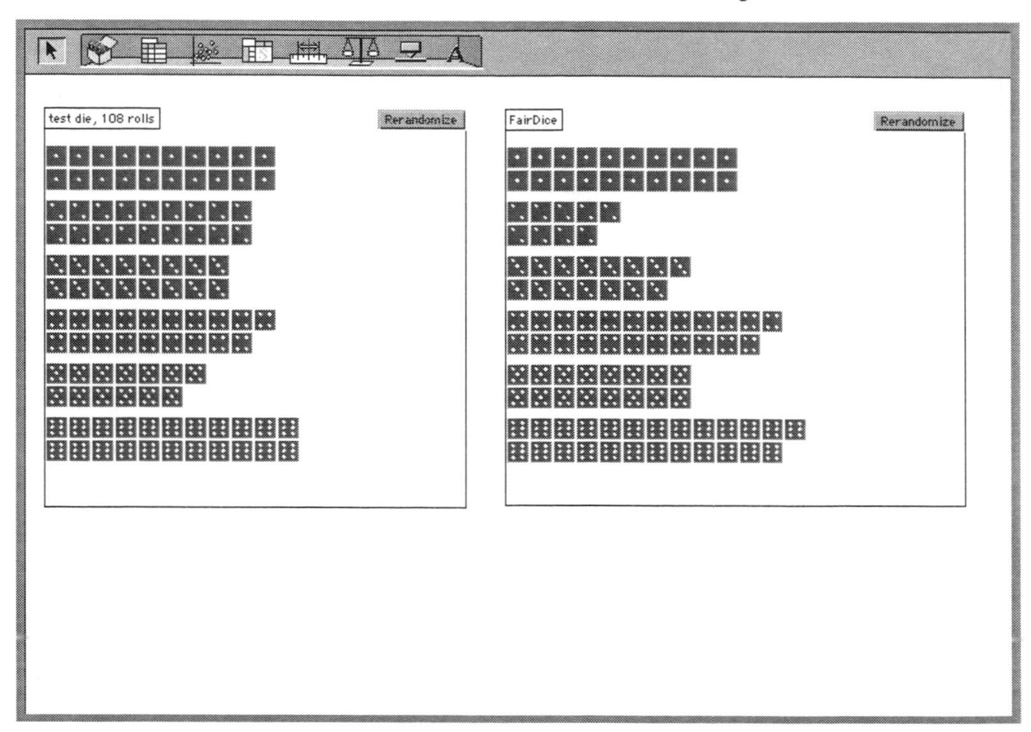

Here you see two collections of 108 cases each. The left one is a test die; the right one is fair. Note in advance that since there are 108 rolls, you would expect 18 of each number (since $6 \times 18 = 108$).

- Press the **Rerandomize** button on the **Fair Dice** (right-hand) collection repeatedly. Note how the distribution is often far from "fair."
- Do the same on the **test die** (left-hand) collection. Again, watch the distribution. This time, see if you can figure out what it is about the distribution that makes it appear unfair. (It is not obvious from the picture above—you'll see it only when you see several examples.)
- Leave the **test die** collection in a "typical" state (the picture above is unusually even; pick one that shows the unfairness) and **Rerandomize** the right again, paying attention to what it is about the test distribution that is different from the fair. (Note: don't **Rerandomize** the **test die** collection again—we'll be comparing fair dice to *this* distribution from here on.)

In general, the **test die** picture should suggest there are more even numbers in the test collection than odd numbers. Now we need to develop a *measure* of how unfair it is. This measure is a *statistic*—a number—that applies to the entire collection. Ideally, it should be large if the distribution is extremely unfair, and small—or zero—if it is fair.

We have created a measure that fits these criteria: **evenMinusOdd**, which is the number of evens minus the number of odds. Let's see what that quantity is.

- Double-click in the left-hand (**test die**) collection to open its inspector. Click on the **Measures** tab to open that panel. It will resemble the inspector in the illustration.

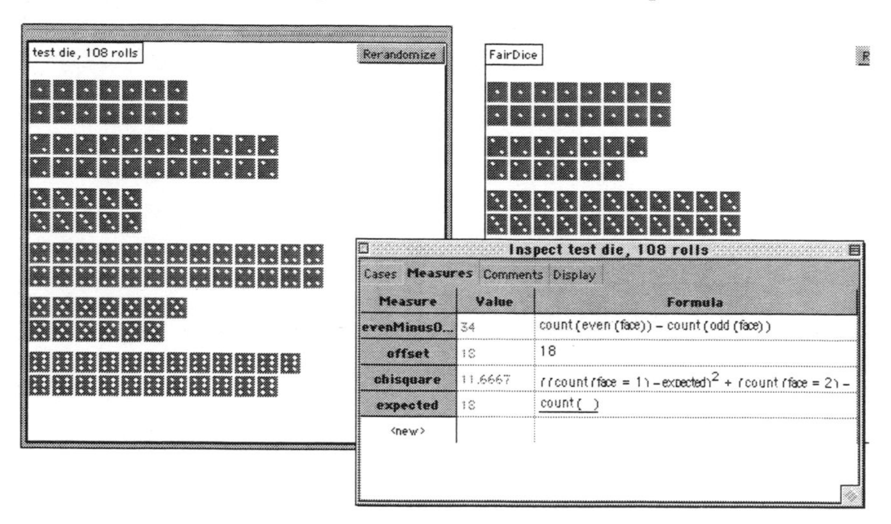

- Note that our value for **evenMinusOdd** is 34 in the illustration. There are 34 more even-numbered dice than odd ones. If everything were flat, that number would be zero.
- Close the inspector. (You might recognize **chiSquare** as another statistic that measures fairness. We'll come back to it later.)

But is 34 (or whatever you have) an *unusual* value? To find out, we will calculate the value for the fair dice. It will not always be zero, because of random variation. Sometimes it could even be quite large. As big as 34? Let's see.

- Double-click the **Fair Dice** collection to open its inspector, and click on the **Measures** tab to go to that panel. There you can see the current value for **evenMinusOdd**. Note it informally (that is, is it bigger than 34?).
- Press the **Rerandomize** button on that collection repeatedly, watching the value of **evenMinusOdd** update in the foreground. Repeat until you have a pretty good sense of whether 34 is unusual. (It is.)

Wouldn't it be great if Fathom would just keep track of these numbers?

- Close the inspector to save screen space.
- Now drag the lower-right corner of the **FairDice** collection up and to the left, to make it so small that it turns into an icon. You will reveal the **Measures from FairDice** collection and a graph of 200 values of **evenMinusOdd**, each one representing a new resampling of 100 fair dice. It should look something like the illustration.

- Let's collect a different set of 200 measures. Select the **Measures from FairDice** collection and choose **Collect More Measures** from the **Analyze** menu. Repeat as many times as you like.

At last we can answer the question, "Is 34 (or whatever number you had) an unusual value?" Look at the distribution of **evenMinusOdd** values in the graph, and see where your test value (34 for us) lies. Chances are, if it's not outside the distribution entirely (as in the illustration above) there are only a few points at 34 or higher.

Since there are 200 points in the graph, you can compute the *P*-value empirically—the probability that a fair die will give a value of **evenMinusOdd** that large *or larger* in a sample of 108 rolls. Generally, if $P < 0.05$, (i.e., fewer than 10 as-extreme points in 200 samples) people consider that significant. That is, a fair die will rarely show behavior like your test die. We can infer that the test die is unfair (and risk being wrong at the 5% level).

Questions

1. Values for **evenMinusOdd** are always even. Why is that?
2. web What is the largest possible value you could get for **evenMinusOdd**? Explain why.
3. What is your *P*-value? **Collect More Measures** to compute several *P*-values.

Extension: chiSquare

The measure **evenMinusOdd** is elegant, but useful only if we observe that the evens outnumber the odds, so that their difference is relevant. There is another statistic, chiSquare (written χ^2), that measures how far the distribution is from "even" no matter how it differs. Its formula is

$$\chi^2 = \sum \frac{(O-E)^2}{E},$$

where the sum is over all possible outcomes—in this case, the die faces 1 through 6. O is the number of cases *observed* to have that outcome, and E is the number *expected* to have that outcome. In our case, E is 18—1/6 of the 108 dice—for each of the six outcomes.

- Open up the inspector for the test die again, and look at the **Measures** panel. Note the value for **chiSquare**. In our example above, it was 11.6667. Close the inspector for screen space.

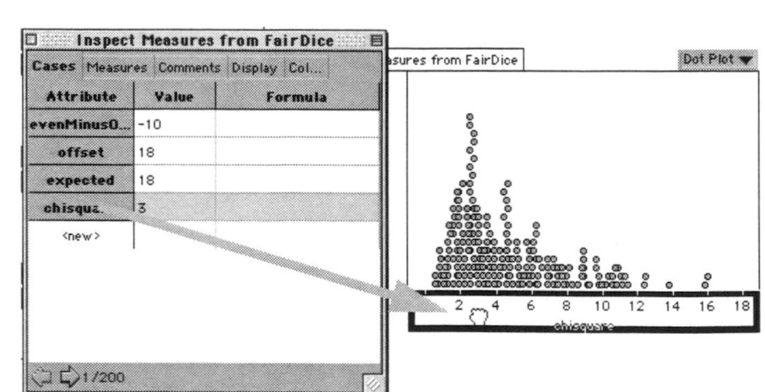

- Open the inspector for the **Measures from FairDice** collection (double-click the box of balls). Click on the **Cases** tab to bring that panel to the front.
- Drag the name of the attribute **chiSquare** to the horizontal axis of the graph, replacing **evenMinusOdd**. Your graph should look something like the one in the illustration.

Note: the attribute **chiSquare** also appears as a measure in the **FairDice** collection itself. If you drag it to a graph, nothing will happen. Be sure you're looking at the **Cases** panel of the **Measures from FairDice** collection.

- Count (approximately) how many cases are equal to or greater than your test value (in our case, where $\chi^2 = 11.6667$, about 8 cases). Compute the *P*-value (for us it is 8/200 = 0.04).
- Collect new sets of measures as before: select the **Measures from FairDice** collection and then choose **Collect More Measures** from the **Analyze** menu.

As before, you can calculate *P*-values repeatedly. In each graph, look at how many sets of 108 fair dice produce results that are at least as extreme as your test distribution.

More Questions

4 In general, does it seem that the test die results are consistent with its being a fair die based on the chiSquare distribution?

web

5 Does the **chiSquare** statistic seem more unusual, less unusual, or about the same as the **evenMinusOdd** statistic?

6 When you compute *P*, why do you divide by 200, even though there are only 108 fair dice in the collection?

Harder Stuff

1 Suppose all of the rolls in 108 die rolls came up *one*. What would chiSquare be then?

2 Explain why, in the equation for χ^2, the measure will be small when the results are "fair" and large when they are grossly unfair.

3 Explain why χ^2 can never be negative.

4 Why do you suppose (as is likely) there were more cases of fair dice having an extreme **chiSquare** than there were of fair dice having an extreme **evenMinusOdd**—that is, a value for the statistic greater than that of the test die?

5 Suppose we had a test value for **evenMinusOdd** of 30. When we look at the sampling distribution, and count how many of the **FairDice** samples are greater than or equal to 30, should we also add in the samples that have an **evenMinusOdd** less than or equal to –30? Explain.

6 Change the unfairness of the test die. You can find it in the formula for **face** in the test die collection. Then see how well you can detect the unfairness of the die.

7 Make a wholly new measure of unfairness and see if it works. You'll need to create a measure for it in both the **test die** (to compute the test statistic) and **FairDice** collections. Then **Collect More Measures** to get a graph of its distribution. Finally, compare the test statistic to the distribution to see how extreme your test die is. One example of such a measure might be "eighteen minus the number of ones."

8 Explore and explain the advantages and disadvantages of using the orthodox measure, χ^2, as opposed to these measures we're just making up ourselves. You can get one insight by doubling the number of cases in the test collection, and then also doubling the number in **FairDice**. You will see that for some statistics, the distribution of sample statistics will change (location, probably not shape) whereas for χ^2, it will not.

Demo 37: Scrambling to Compare Means

Randomization test • Using scrambling to simulate the null hypothesis • Generating a sampling distribution

Suppose we want to test a new fertilizer. We make an experiment in which we assign aspidistras to two groups randomly: the experimental group gets the new fertilizer and the control group gets the old one. Four weeks later, we measure the plants, and compare the two groups. If the average experimental aspidistra is taller, the fertilizer is good, right?

Alas, it's not so easy. Even if we treated the plants identically, there would be some variation, and if we assigned the plants to different groups—and treated them the same—one group would be taller on the average than the other. So the question becomes, if the fertilizer didn't do anything (we call this the *null hypothesis*), how likely is it that the intergroup difference we measured (or more) would arise by chance? If it is very unlikely, we conclude that the fertilizer did indeed make a difference. Often, researchers define "very unlikely" as being a probability of less than 5%, but, as with so many things in statistics, you can decide for yourself what probability you want to use.

In this demo, we'll test the fertilizer using *scrambling*—officially known as a randomization test. The whole idea is to simulate the null hypothesis, that is, to alter the data so that any difference between the experimental group and the control is due only to chance. We'll do this repeatedly to make a *sampling distribution* of the statistic we need to look at; we compare our *test statistic* to the sampling distribution to decide whether to reject the null hypothesis.

What To Do

- Open the file **Scrambling.ftm**. It should look like this:

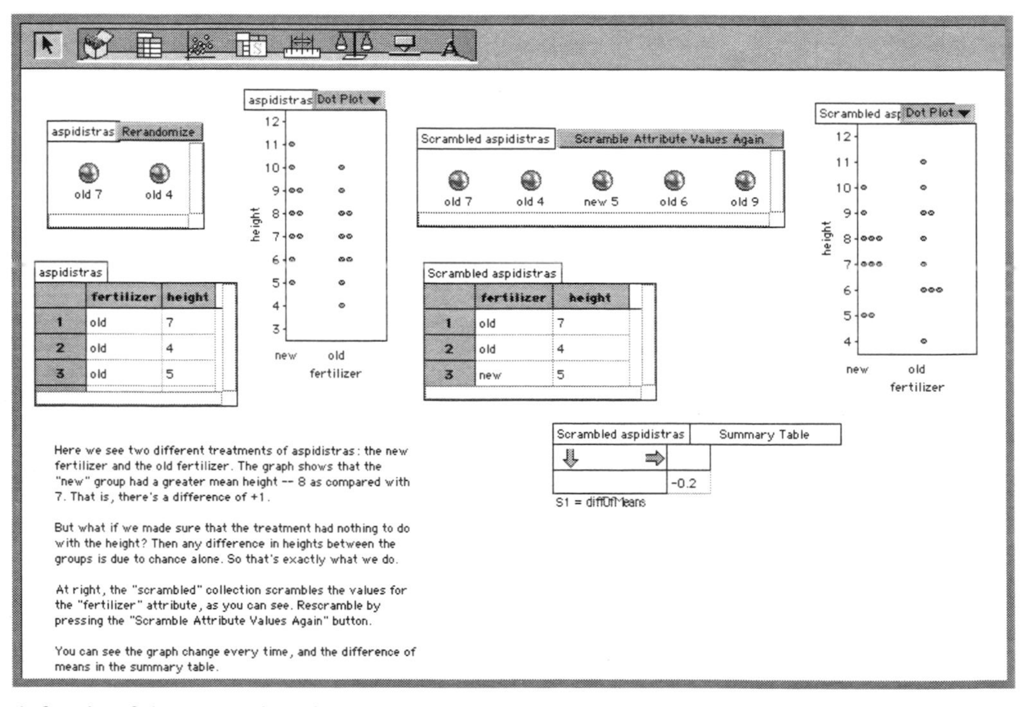

The left side of the screen has the original (made-up) data and a graph showing the difference between the two groups. The right-hand side shows the data with the **fertilizer** attribute scrambled. You can see this by comparing the gold balls or the case tables.

In the summary table, you can see a value for **diffOfMeans**. For this case, it's –0.2. This is the difference between the means of the experimental and control groups *in the scrambled collection*. In our original data, the difference of means is the *test statistic*, and it has a value of +1.0. Speaking symbolically, the formula is:

diffOfMeans= mean(height,fertilizer = "new") – mean(height, fertilizer = "old")

- Press the **Scramble Attribute Values Again** button and see what happens. Do so repeatedly.

Questions

1. How can you tell that the **fertilizer** attribute was scrambled and not **height**?
2. Looking informally at the summary table for **diffOfMeans**, what's the range of values you get for the scrambled statistic?
3. Is our test statistic of +1.0 particularly unusual, or is it the sort of thing that arises by chance? (web)

Moving On

We looked at those **diffOfMeans**s informally. Let's use Fathom to record those values and look more closely at the distribution of that statistic.

- Open the file **Scrambling 2.ftm**. It should look like this:

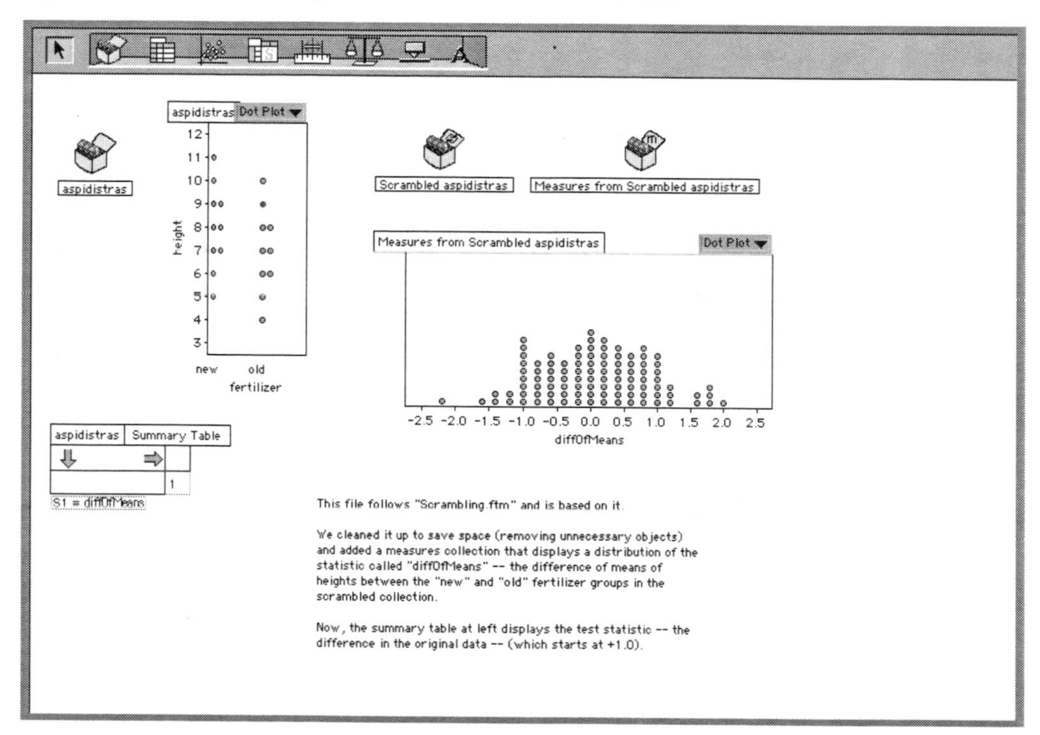

This file compresses elements of the previous one and adds some "measures" machinery. The collection **Measures from Scrambled aspidistras** contains 100 values from rescramblings, which are displayed in the graph. Note that the summary table here refers to the original data, not the scrambled collection. Thus, it shows the test statistic (+1.0) that we compare to the sampling distribution in the graph.

- Count how many cases in the distribution have values for **diffOfMeans** that are +1.0 or higher.
- Re-collect measures: click the measures collection once to select it, then choose **Collect More Measures** from the **Analyze** menu. Fathom will collect 100 more measures, rescrambling the scrambled collection every time, and recomputing the statistic (that's why it takes a while). See how many are +1.0 or above again. Repeat as many times as you need to in order to understand what's going on.

- Estimate the probability that a scrambled collection has a **diffOfMeans** of +1.0 or more. Is it more than 5%?

It seems as though the two groups just aren't different enough in the mean to be distinguished from chance. Let's change the original data and see if we can make it less likely for our test statistic to arise by chance:

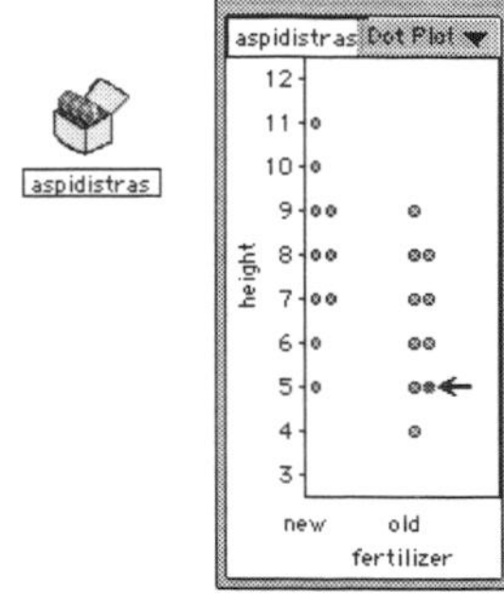

- In the graph of the original data, drag the top case in the **old** fertilizer group (value of **height = 10**) down to a value of 5, as shown. The test statistic in the summary table should increase (we're increasing the difference of means) from +1.0 to +1.5.
- Again re-collect measures. Select the measures collection and choose **Collect More Measures** from the **Analyze** menu.
- Now count: how many cases equal or exceed the test statistic? Note: the test statistic is now 1.5, not 1.0—so there are fewer cases than you might think.

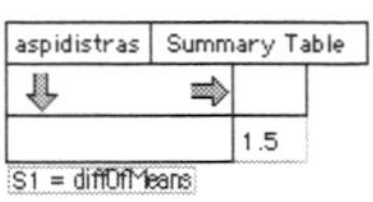

You'll probably find that it is now less likely—but still possible—to get a **diffOfMeans** as large as 1.5. Let's make it even less likely.

- Repeat the above experiment, gradually altering data to make the groups more different. You might lower the same point, or raise a point in the other group. Continue until the test statistic seldom occurs.

Harder Stuff

4. In this randomization test, the *P*-value is the probability that a scrambled **diffOfMeans** is greater than or equal to the test statistic. So if there were 7 cases out of 100 at +1.0 or greater, the *P*-value would be 0.07. At what *P*-value are you convinced that the group means really are different?

web

5. What result would you get if the original data had disjoint groups—that is, the highest value for one group was lower than the lowest value for the other?
6. Explain why the logic is always "greater than or equal to" the test statistic. Why not just greater than? Why not just equal to?
7. Explain the logic of this test. What is the test statistic? Why do we scramble? What is that distribution a distribution *of?* What does it mean when we compare the test statistic to the distribution? What possible results can there be?
8. Decide whether this test should really be one-tailed or two-tailed, and explain your reasoning. That is, when we compare a test statistic of +1.0 to a distribution of **diffOfMeans**s, should we count the cases where **diffOfMeans** is greater than +1.0, or where the *absolute value* of **diffOfMeans** is greater than +1.0?

What You Should Take Away

Scrambling is genuinely useful for doing statistical inference—in this case, deciding whether the two means are different—especially if the data don't fit orthodox requirements of normality. But more important for learning is that scrambling can show you what the test is really all about: comparing a test statistic to a sampling distribution. And that distribution comes from a process (scrambling) that guarantees that the null hypothesis is true.

Demo 38: Using a t-Test to Compare Means

Comparing means with Student's t

In "Scrambling to Compare Means" on page 126, we used randomization to compare the means of two groups. Now we'll compare the same two groups using the orthodox *t*-test for difference of means.

What To Do

- Open the file **Compare Means Using t.ftm**. It will look like this:

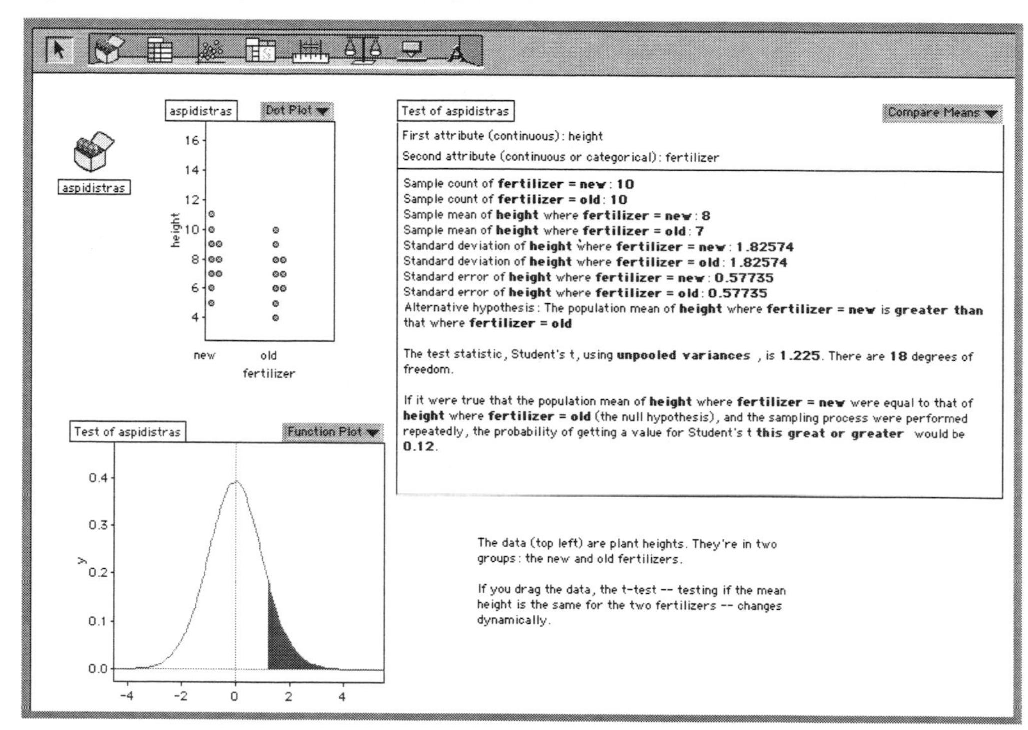

This shows the same data as before, but with Fathom's built-in *t*-test to help us decide if the two groups are different. The *P*-value (at the end of the test object) is 0.12—not usually small enough to justify rejecting the null hypothesis. Let's change the data.

- Play with the data to see how changing a single point affects the results of the test.
- Arrange the data, changing only one data point, so that the *P*-value is 0.05 or lower.

Questions

web

1. How can you predict, before you drag a given point, which way the *P*-value will change?
2. At the bottom of the window, what does the graph of the curve show?

Extension

- Return the data (through re-opening the file or by multiple undo) to the way it was in the beginning.
- In the graph of the data, grab the lowest point among the **new** fertilizer data (**height = 5**) and slowly drag it upward. This will make the two groups more different.
- Watch the *P*-value as you drag. It goes down (to about 0.028) and then, eventually, starts back up.

Harder Stuff

web

3 Explain how that is possible. That is, by dragging a point from the higher group *up*, you're making the two groups more different, aren't you? So it makes sense that *p* should decrease. But then why does it increase again?

4 Investigate the two-tailed case: first, figure out how to change the test to a two-tailed test. Then describe how the results and displays are different. How much more (or less) do you have to change the data to get a *P*-value of 0.05 or less?

5 If you have done "Paired Versus Unpaired" on page 135 (or even if you haven't), why do you suppose we didn't use a paired test here? Aren't paired tests generally more effective?

The Point

Here are two things you should take away from this demo:

- We often think of the test as comparing the means (in fact, it's called Compare Means, isn't it?). But the whole thing hinges on the *spreads* in the groups. Why? Because they set the scale by which the difference is measured: that difference is not measured in inches or centimeters—it's measured in standard errors.
- The results of a hypothesis test can be very sensitive to individual data points. You can see in this demo that one wacky point—or even a not-so-wacky one if your sample is small—can mean the difference between a highly-significant *P*-value and a boring one. And it can go either way. Therefore, be ye skeptical: *Always* look at the data. Make the graph. Play "what-if" games with the points. If you just turn the crank on the test and cheer or weep depending on how *P* compares to 0.05, you're rushing things.

Demo 39: Another Look at a t-Test

Repeated t-tests on samples from the same distribution • How t, P, mean, and standard deviation interrelate

In this demo, we'll perform a *t*-test on the mean of a distribution that we have constructed ourselves. We'll know whether the null hypothesis is true. But then we'll change the distribution and see how the test changes, recording the results of these many *t*-tests. Throughout the demo, we'll look at those results as data to see what we can learn about the process.

What To Do

The specific numbers in your file will be different because of the randomization.

- Open the file **Exploring t.ftm**. It should look something like this:

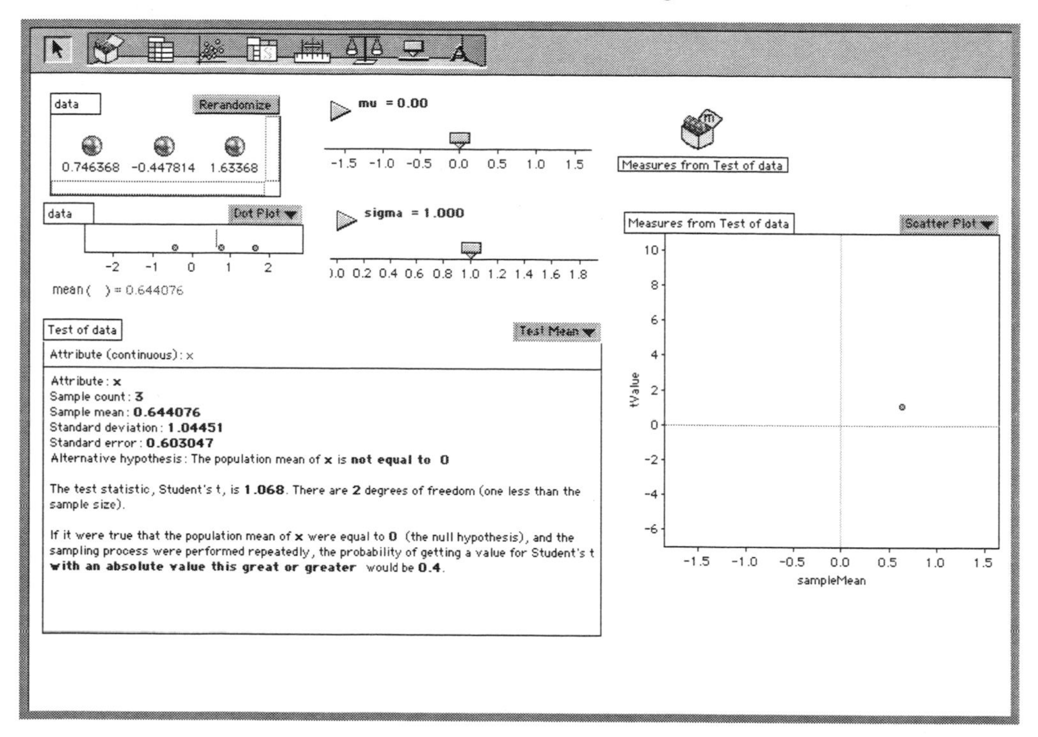

You can see the data in the upper left. Fathom draws the three values from a normal distribution controlled by the sliders **mu** and **sigma**, which determine the population mean and standard deviation, respectively. Below that is a graph of the data, and then a *t*-test, testing the null hypothesis that the mean is equal to zero. In the illustration, see how the sample mean is close to zero, *t* is small, and *p* is large? We cannot reject the null hypothesis, which is not a bad thing, especially since the null hypothesis is *true*: the simulation *sets* the population mean to zero!

- What else could have happened? Click the **Rerandomize** button a few times. Fathom redoes the *t*-test, and also plots the sample mean and the *t*-value in the graph on the lower right.

You probably see a swath of points with a roughly positive slope. This makes sense because *t* is proportional to the sample mean (since we're testing against zero). But they aren't lined up perfectly, because *t* also depends on the sample standard deviation, which varies from run to run.

- Let's look at the standard deviation. Grab the lower-right corner of the test (the one with all the text) and drag it up and to the left until it turns into an icon. This reveals a graph of sample standard deviation versus sample mean.
- As shown in the illustration, drag a long horizontal rectangle in the new graph to select all the points with the smallest standard deviations.
- Notice where these points lie on the other graph; select other groups of points and figure out why they are where they are on each graph.

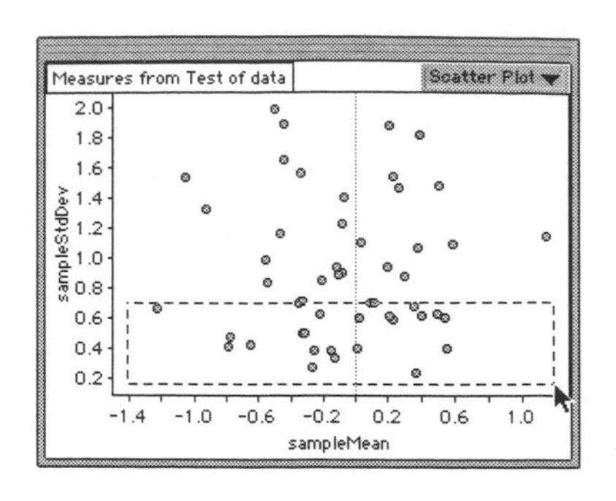

Note: there are at least two ways to "explain" such relationships. One is to relate them to the formula: standard deviation is in the denominator for t, so a large SD means a small t. The other is to relate them to meaning: if a sample has a large SD, we're less sure about our estimate for the mean. So the bigger the SD, the smaller the significance of any difference between the sample mean and what we're testing.

Questions

web 1 In the first graph, why are there points only in the first and third quadrants? Is it possible to have points anywhere in those quadrants?

web 2 How can it be that one sample with a mean of 0.6 can have a smaller (i.e., less significant) value for t than a different sample with a mean of 0.4?

Onward!

Do you think one of these many t-tests rejects the null hypothesis? That would be a Type I error, since the null hypothesis is true. Let's find out.

- On the right-hand graph, double-click the point with the largest (absolute) value for t. The collection's inspector opens to show that point. See the attribute's **pValue**. (Those attributes tell you the results of the test when it was performed; we have set up Fathom to collect these results automatically.)

Was is less than 0.05? Maybe, maybe not. Let's see some examples of tests where we would definitely reject the null hypothesis. A great way to do that is to make it false.

- Close the inspector.
- Drag the **mu** slider back and forth; Fathom collects points as fast as it can. You do not need to go any farther than ±1.5 to see everything.
- Now inspect the point with the most extreme t. Find its **pValue**.
- Let's look at **pValue** directly in the graphs. Drag its name from the inspector to the *middle* of each graph. It will turn colored, and you will see a spectrum-legend at the bottom.

More Questions

3 Why does the **pValue** have the range it does?

4 Look at the regions on each graph where **pValue** is near one (they're probably colored red). Why are those regions where they are?

web 5 Can you tell from a point on the graph whether it was made with **mu = 0** or not?

Demo 40: On the Equivalence of Tests and Estimates

How a hypothesis test and a confidence interval are really the same

In ancient times, statistical tables gave only a critical value—the value for a statistic at the edge of significance for some significance level, usually 5%. If your t or χ^2 was bigger than the critical value, your result was significant. When you tabulated your research results in a journal, you placed an asterisk next to each one that was significant.

Unfortunately, the reader had no way of knowing *how* significant it was (unless you put in lots of asterisks).

Nowadays, we can calculate the actual *P*-values easily, or give confidence intervals; that way readers can make up their own minds whether the result is significant to their satisfaction.

But which is better—the confidence interval or the *P*-value? This demo shows that, in one sense at least, they are equivalent. Suppose you're testing whether the mean of a population is zero. You construct 95% confidence intervals of the mean. You also perform a *t*-test, and set the significance level—the critical *P*-value α—at 0.05. Then a significant result, i.e., $P < \alpha$, occurs whenever the confidence interval does not contain zero, and vice versa.

What To Do

- Open the file **Tests and CIs.ftm**. It will look something like this:

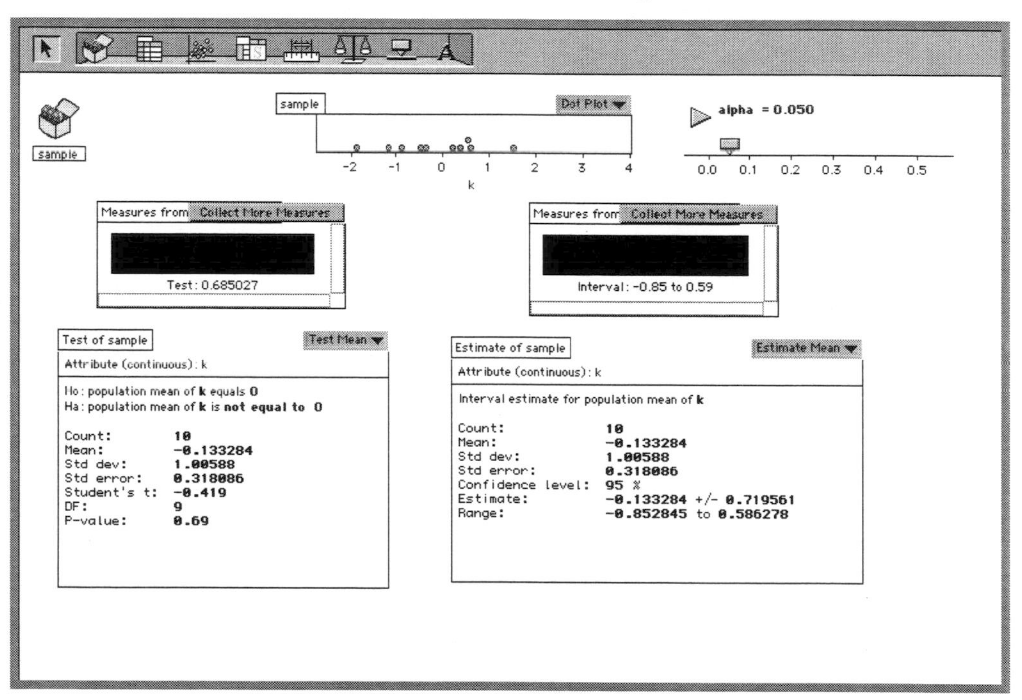

At the center top, you see our data, called **sample**, displayed in a dot plot. Below the data are two mysterious black bars. Below them are, on the left, a hypothesis test that the mean is not equal to zero, and on the right, an estimate showing the confidence interval for the mean. At the upper right, a slider controls the significance level **alpha**.

Now to the black bars: they show the test (left) and estimate (right) results. The size of the bars doesn't matter. But they change to red if the test shows significance, either because $P < \alpha$, or because the confidence interval no longer contains zero.

- Try it! Drag points in the graph to make it clear that the mean of the population is not zero. As you pass the critical values, note that the two bars change color at the same time. Repeat as necessary.

- Move the points so that the rectangles just barely turn from red to black.
- Now drag the **alpha** slider so that the rectangles go red again. (If, while dragging **alpha**, one of the black boxes turns into a giant squashed gold ball, don't be alarmed.)
- Play with both effects as long as you like!

Questions

1. Which direction do you have to move the **alpha** slider to make black rectangles red? Why does that make sense?
2. Which way to make red rectangles black?
3. As you move the **alpha** slider, the number under the left-hand box does not change. Why not?
4. [web] Why does the right-hand rectangle go nuts (e.g., turn into a fat gold ball) sometimes?
5. [web] Why doesn't the left-hand rectangle go nuts at the same time?
6. What happens when the sample mean is zero? (Note: you will probably have to make a case table to get it to be exactly zero.)

Harder Stuff

7. Based on what confidence intervals, the null hypothesis, and hypothesis tests *mean*, explain why it makes sense that this demo should work.
8. Make a case that one or the other inferential tool—the hypothesis test or the confidence interval—is better, even though they are equivalent in this sense of this demo.

Demo 41: Paired Versus Unpaired

How a paired test gives a significant result more easily than its unpaired counterpart

Suppose you've given a pre-test and a post-test to a group of students and you want to see if your instruction had any effect. In your statistical armamentarium for comparing two groups, you can test using either *paired* or *unpaired* techniques. In a paired test, you take the two scores for each individual student, and find the difference; that is, you see how much each student's score changed. In the unpaired test, you test to see whether the "after" mean is significantly higher than the "before."

Which should you use? In this situation, paired is best, because it's more powerful: you get a smaller *P*-value with the same data. Let's see how it looks...

What To Do

- Open the file **Paired Versus Unpaired.ftm**. It will look something like this:

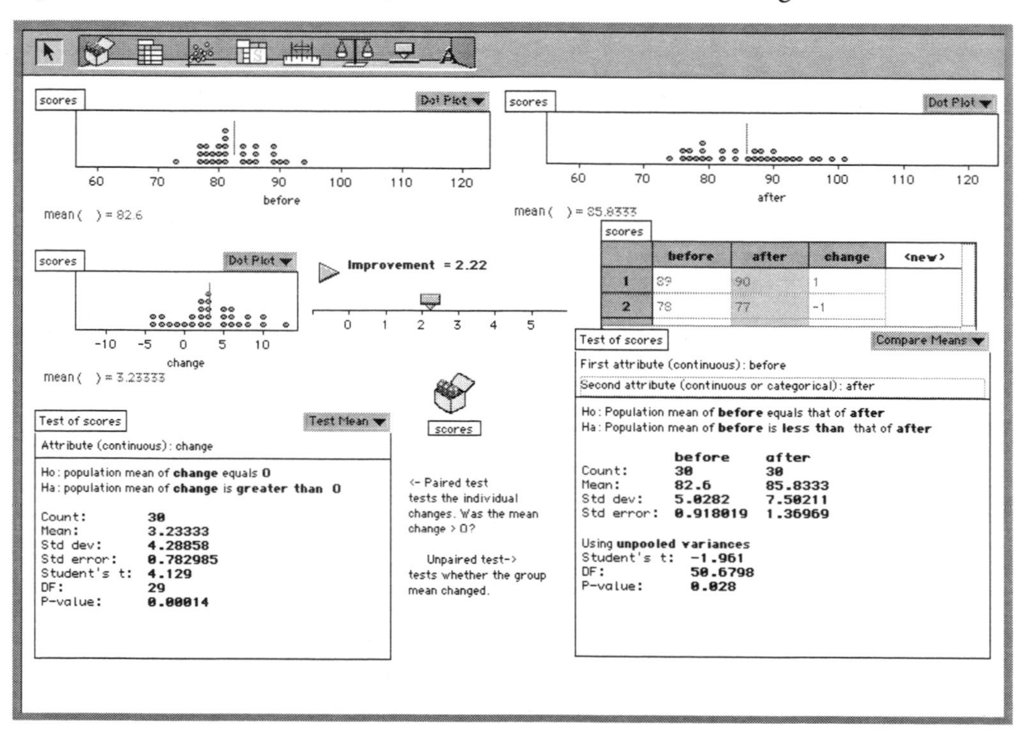

This is a complex-looking document, but there's really not a lot to it. You see the top of the data in the case table in the middle on the right. In it, you can see three attributes: **before**, **after**, and **change**, representing the pre- and post- scores and the individual change in score. There's also a slider, **Improvement**, that controls the average amount of that change.

Then, in the lower left, is the "paired" test: we test whether the mean of **change** is greater than zero. Lower right, below the case table, you see the "unpaired" test: we compare means to see if the mean of **after** is greater than that of **before**.

- Click on a point in any of the three dot plots. The same case will highlight in the other two graphs. Click on other points, too; study them to make sure you understand how the three attributes are related.

- Observe the test results. In the illustration, for example, the *P*-value for the paired (left-hand) test is smaller—more significant—than that of the unpaired (right-hand) test.

The shortcut for **Rerandomize** is **clover-Y** on the Mac or **control-Y** in Windows.

- Choose **Rerandomize** from the **Analyze** menu. The points and the test results change. Again observe the difference in the *P*-values.
- Move the **Improvement** slider, observing the effect on the graphs and on the test results.
- Find a value for the slider **Improvement** so that the *P*-value of the paired test is generally below 5%, while the unpaired test is generally above. (**Improvement = 1.5** is not bad.) **Rerandomize** repeatedly to get a sense of the variation.

At this point, you've been focusing on the numbers. Now step back and look at the graphs. The question is, can you see the difference between the two sides? If so, you would see this: it's fairly clear that the left-hand **change** graph does not come from a population with a mean of zero (that is, the individual scores have generally improved) but it's not clear that the two right-hand graphs—**before** and **after**—come from populations with different means.

You can also think of it this way: if your **before** data are all spread out, but *every* student makes a small improvement, the mean of **after** will be higher, but not by much compared with the spread in the distribution of scores. So the unpaired test will have a small *t*-value. On the other hand, **change** will be tightly clustered around a number higher than zero—a number larger than its spread. So the paired test gets a large value for Student's *t*, and a lower *P*-value. The paired test will be significant; whether it's meaningful is a different issue!

- Finally, once again select single points on the graphs to reinforce your understanding of their connection.

Questions

1. The slider **Improvement** should be roughly equal to the mean improvement of the scores. Is it?
2. Where can you find the mean improvement in the scores in the paired test?
3. How about in the unpaired test? (web)

Harder Stuff

4. Explain in words, as clearly as possible, how it can be that a paired test would give you better results than an unpaired test.
5. For the unpaired test, we compared two means. That makes sense. But for the paired test, we tested only a single mean. Explain why.
6. The formula for **before** is **randomBeta(4, 4, 70,100)**. Change the formula for **after** to be **randomBeta(4, 4, 70 + Improvement, 100 + Improvement)**. Now see what happens and explain it.
7. Suppose you're designing a study to see whether students' attitudes towards drug use change after they have seen a series of films. Your main measurement instrument is a questionnaire. What are the main arguments for and against having students put their names on it? (web)

Demo 42: Analysis of Variance

Assessing whether means are different in different groups • Introduction to ANOVA

This is an advanced topic for an introductory course, and the file for this demo is complicated, but the basic idea is simple. Suppose you're growing plants. Some are in the **sun** and some are in the **shade**, and you measure their **height**. The height is a *response* variable; we wonder if it depends on the *treatments*, **sun** and **shade**. We could use a *t*-test to see if the mean of the **sun** groups was significantly different from the **shade** group. But what if there were *three* groups instead of two? How would you calculate *t*?

We need to define a new measure for how different the groups are. We'll compare the new measure's test value to its distribution under the null hypothesis—where there is no dependence at all. This is exactly what we've done with Student's *t* and chiSquare and any other statistic we use to construct a test. We'll construct the measure using the same reasoning we used for *t*. There, we measured the difference *between* the groups in units of standard errors—in units of the variation *within* the groups. Generalizing this idea to more than two groups is a little tricky, but the idea is the same: we figure out how much of the spread is due to variation *within* the groups and how much is due to variation *between* the groups, and we make a ratio:

$$\textit{dependence} = \frac{\textit{variation between groups}}{\textit{variation within groups}}$$

In that way, a large dependence will yield a big number—when there is little variation within groups compared to the variation between. And if the groups are essentially identical, the between-group variation will be small, and our statistics will be close to zero. There are details, of course, but they aren't particularly diabolical. The main thing to bear in mind from the beginning is that we will measure variation using *sums of squares* of deviations from means; that is, we'll look at numbers that are more like variances than like standard deviations.

What To Do

✣ Open the file **Within and Between.ftm**. It will look like this:

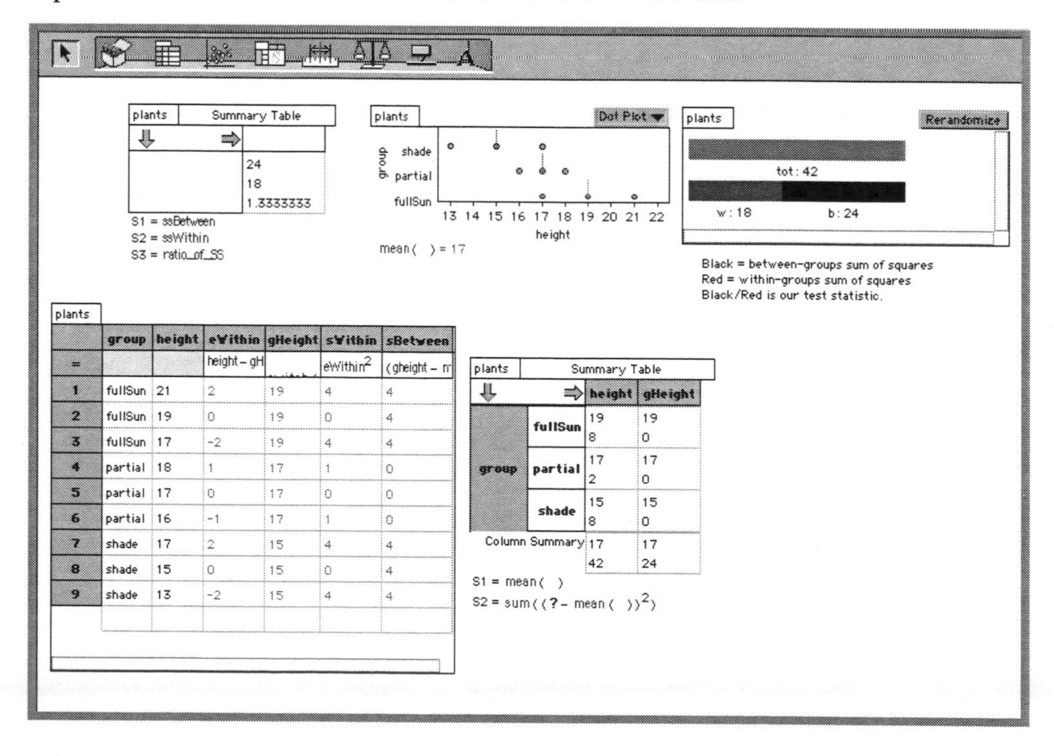

This is all about (imaginary) plant data. The case table in the lower left is a good place to start: there you see the raw data—the **group** (**fullSun**, **shade**, or **partial**) and **height** of each of nine plants. You also see three more attributes, calculated from the original two:

- **gHeight** is the group height—the mean of the heights of all the plants in the group. This has the same value for every plant in that group, as you can see.
- **eWithin** is the "error" of the plant's height "within the group." That is, you can think of every **height** as being **gHeight + eWithin**.
- **sWithin** is **eWithin** squared.
- **sBetween** is the squared residual of that plant's *group* (not the individual plant) within the whole data set. That is, it's the square of **gHeight – mean(height)**. This will define what we mean by *between-group variation*: it's the variance of the set of groups, weighted by the number of cases in each group.

Directly above the case table is a summary table showing the sum of the squares of the within-group residuals (**ssWithin**) which is simply the sum of all the numbers in the **sWithin** column. Similarly for **ssBetween**. Then **ratio_of_SS** is the ratio we talked about earlier, computed by taking **ssBetween / ssWithin**. So the value of this "dependence" measure here is 1.33; we don't know yet if that is a lot or not.

To the right is a graph showing all the data; below the graph is another summary table, this time showing statistics for the **height** and **gHeight** attributes for the groups: the mean values, and the sum of the squares of their residuals *in the context of the cell they're in*. This is a subtle but important table. At the lower left, you can see the total sum of squares of the residuals (42) and the grand mean (17). Within the left column, you see the sums of squares of the residuals within each group (8, 2, and 8, for a total of 18). Finally, at the lower right—in the column summary for **gHeight**—is yet another sum of the squares of the residuals. This time, each residual is the distance of the mean of the *group* from the total mean; so this is the *between*-groups sum of squares, 24. These numbers demonstrate an important identity:

$$SS_{total} = SS_{within} + SS_{between}$$

since 42 = 18 + 24. In the upper right is a display that shows how the total sum of squares (the green bar) is divided into the within and between parts (red and black).

- In the graph at the top, grab any point and drag it. Study its effect on the displays and numbers. Undo and try it with a different point. See if you can understand why things change the way they do.
- Now try to grab a point and move it in a direction that will make the differences among the groups more pronounced. (For example, move the **fullSun** point at **height = 17** to the right.) Verify that the **ratio_of_SS** value increases (see the upper-left summary table) and that in the bar-display, there is more and more black compared with red.
- Repeat, exploring how the relationships among the groups affect the **ratio_of_SS** statistic and the size of its two components, **ssWithin** and **ssBetween**.

Questions

1. In the lower summary table, the top number in each box is the same in the left column as in the right. Why?
2. In the lower summary table, the right-hand column has a bunch of zeros in it. Why are those numbers zero? (web)

Onward!

Now let's figure out if our value for **ratio_of_SS** is unusual, or if it could arise easily by chance. We'll use the same strategy we used in "Scrambling to Compare Means" on page 126—we want to see what happens in the case of the null hypothesis. So we will make the null hypothesis true: we will scramble the values in the **group** attribute so that any apparent relationship between **group** and **height** is due to chance alone. Then we compute **ratio_of_SS** for those data, and repeat the process, building up a distribution of the **ratio_of_SS** statistic for the case that the null hypothesis is true.

- Put everything back the way it was.
- To save space, shrink the upper-right "bar" display until it turns into an icon (**plants**). Then move it to the upper-left (empty) corner of the window. Do the same with the text below it.
- Finally, choose **Show Hidden Objects** from the **Display** menu. The file should now look something like this:

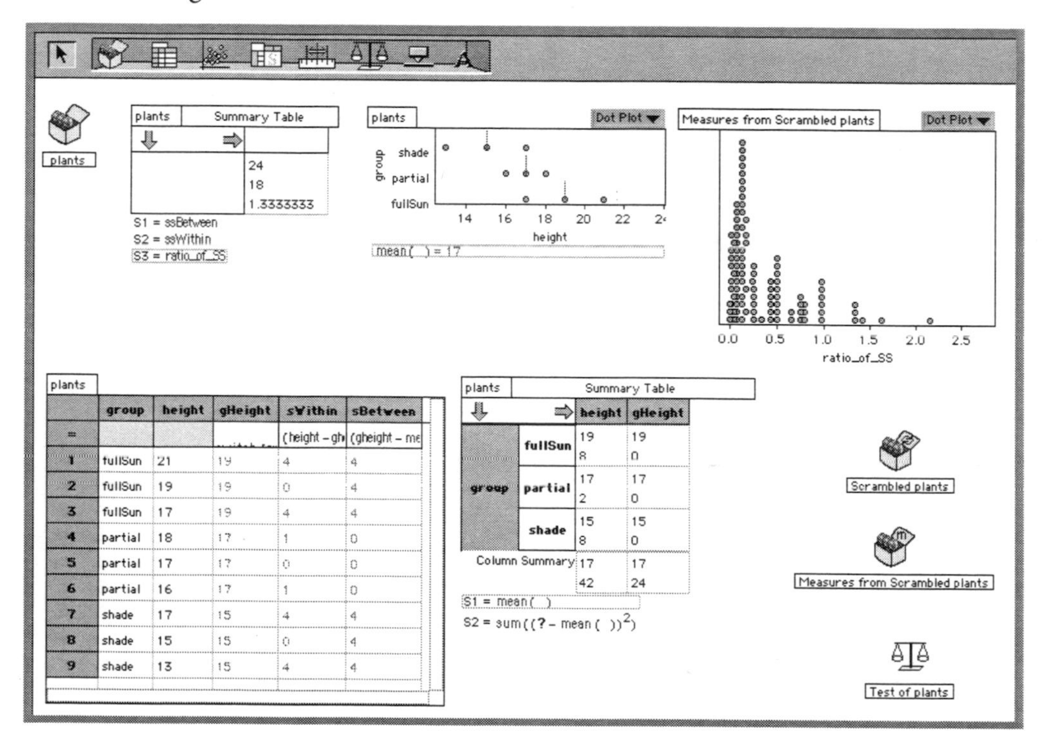

Now you can see the sampling distribution on the right. It has 100 points. You can see that 6 of the 100 points have values of **ratio_of_SS** equal to or greater than our test statistic, 1.33. Thus, 0.06 is our empirical *P*-value.

- Drag data in the **plants** graph to make the statistic bigger and the groups more distinct.
- Select the lower-right collection, **Measures from Scrambled plants**, by clicking on it once. Then choose **Collect More Measures** from the **Analyze** menu. Fathom will churn away and eventually construct a new graph.
- Compare the new test statistic to the new graph; repeat as necessary.

Extension: Using the *F* Statistic

The ratio we used is similar to the orthodox statistic they use in *Analysis of Variance* (ANOVA): the *F* statistic. We used the ratio of sums of squares; the *F* statistic actually looks at the ratio of the variances. To get the variance—the mean square deviation—from the sum of the squares of the deviation, you have to divide by the number of *degrees of freedom*[1] (not *n*, because we are trying to infer from a sample). Let's look at *F* and see what happens:

- Click the top summary table to select it. Then choose **Add Formula** from the **Summary** menu. The formula editor appears.
- Enter the one-letter formula **F**. (We have already defined it as a measure; you may look at its formula if you like; it's a measure in **plants**.) Close the formula editor with **OK**.

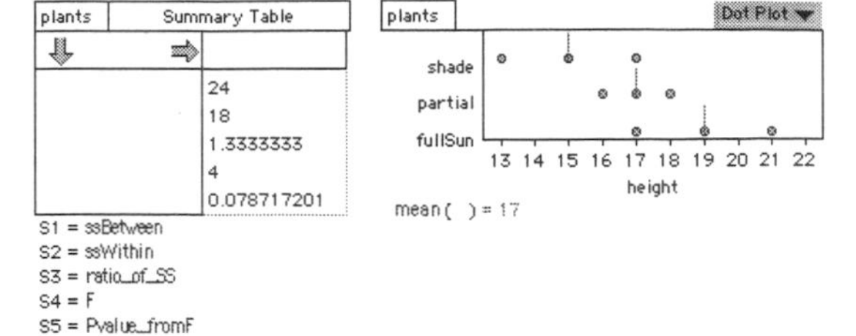

- Now add another formula to the table, simply **pValue_fromF**. This is the *P*-value from an *F*-test. You may need to stretch the summary table to see all of the values; they will look something like the illustration. (Here, *P* is about 0.08; if there were no relationship between **group** and **height**, we'd see this value of *F* or greater 8% of the time.)
- Finally, open up the inspector for the lower-right measures collection, go to the **Cases** panel, and drag **F** to the horizontal axis, replacing **ratio_of_SS**. Now you can compare the test value for **F** with the distribution.
- One more thing: there is a *test* in the lower-right corner. Drag it into the middle and expand it; it's a Fathom ANOVA test, which does all that we have just done in a simple display.

More Questions

3 If the null hypothesis were true, which would usually be bigger—the within-groups or the between-groups variance?

4 What value for **F** gives a *P*-value of about 0.05? (i.e., What is the critical value for *F* at the 0.05 level?)

5 In our example, our three groups' means are more or less evenly spaced. Is the *F* statistic you get larger, smaller, or the same as if you take the one in the middle and move it to the end—so we'd have two groups about the same and one different?

Harder Stuff

6 All we're doing here is comparing means. Why can't we just use a *t* test?

7 We have blithely stated—and shown empirically—that the sum of the squares of the within-group residuals, plus the sum of the squares of the between-group residuals, equals the sum of the squares of the "total" residuals (the distances of each data value from the mean of the entire data set). That is,

$$SS_{total} = SS_{within} + SS_{between}$$

Prove it.

web

8 Plot **F** against **ratio_of_SS**. Explain the graph you see.

---

1. The number of degrees of freedom—*df*—is a lot like the $n - 1$ we divide by to get a sample SD. So since there are 3 groups, the *df* for the "between" is 2. For within-groups, we get *df* of 2 for each group, for a total of 6. Notice that the total *df* is 8—one less than the number of points. We don't treat *df* in this book; if we figure out a great way to do it, we'll post it at **http://www.eeps.com...**

Power in Tests

The null hypothesis is dull; if it's true, nothing is happening. Usually, we're trying to reject the null hypothesis—unless, of course, the null is true.

Yet the null hypothesis is hardly ever really, actually, exactly true. Even if we fail to reject the null hypothesis (e.g., if P is large) that doesn't mean the null *is* true—only that our results would be plausible if it *were*. Are the two populations *exactly* identical? No way. We just may not be able to show that convincingly.

So, given that we believe the null hypothesis is false, what's the chance that we get a significant result from our test? That's what we mean by *power*. And power depends on many factors. One is the sample size: the bigger the sample, the better chance you have of rejecting that null. Another is the test itself: some tests are just more powerful than others. Another is the alpha level of the test: there is a trade-off between false positives and false negatives. And, of course, another factor is reality: how far is the actual population mean from the hypothesized value? A long way? Then you have a better chance.

This section includes:

"The Distribution of P-values" on page 142. If you resample many times, and collect the distribution of P-values from the tests you do, what does that distribution look like? If the null hypothesis is true, that distribution is *flat*.

"Power" on page 144. Power—the chance of rejecting the null—is a function. In this demo, we perform a lot of tests with different true population values. That way, we construct power from empirical probabilities—as a function of the true population value. Sure enough, as we get farther from the null value, the probability approaches 1.

"Power and Sample Size" on page 147. This is the same setup as the previous demo, except we change the sample size instead. This is one way people often use power: if you think some quantity has a mean value of 6 and not 5, you can calculate the sample size you need to prove it.

"Heteroscedasticity and its Consequences" on page 149. This is only a slight detour at the end of this power-trip. Inference for regression assumes that the variance of the dependent variable is the same everywhere, that is, it's *homoscedastic*. This demo shows what happens when it's not.

web

Remember: If you see web in the side margin, that means that there's a solution on the web at the time of publication. See **http://www.eeps.com**.

Demo 43: The Distribution of P-values

How the distribution of P is flat if the null hypothesis is true • How it changes if the null hypothesis is false

This demo assumes that you know about hypothesis tests and what *P*-values are. You can think of it as an extension of "Another Look at a t-Test" on page 131; in fact, the file uses a lot of the same machinery. Here we focus on the distribution of *P*-values, which leads to discussions of power.

- Open the file **Distribution of P-values.ftm**. It should look something like this:

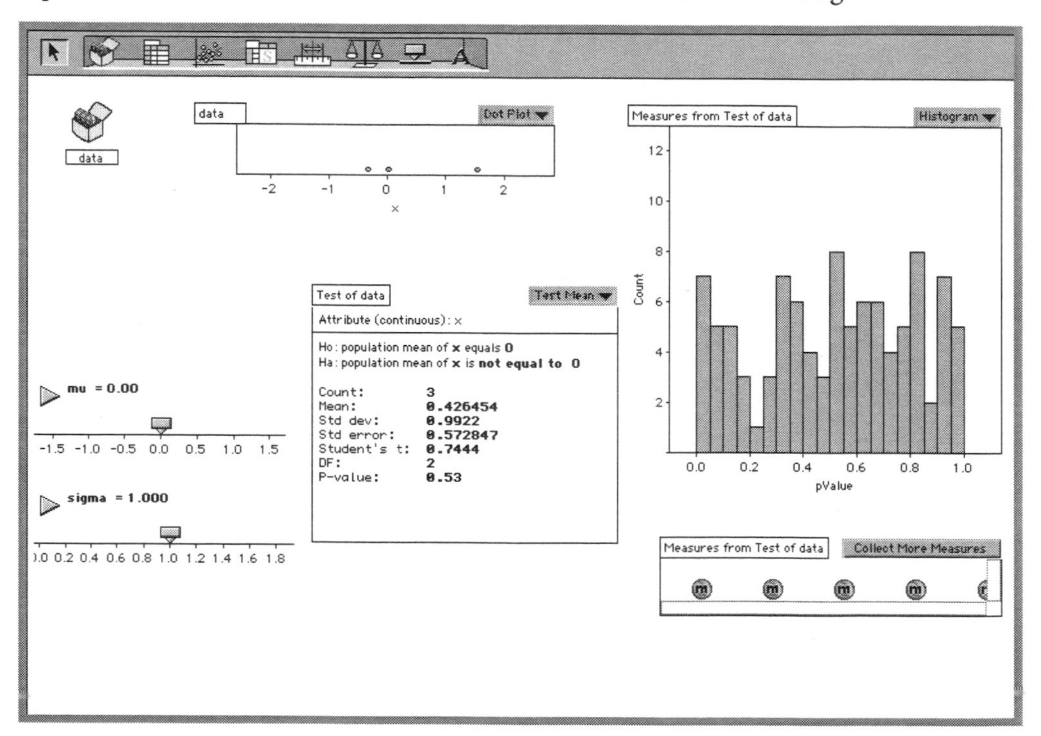

In this document, the data—the collection in the upper left—are once again three points drawn from a normal distribution with mean **mu** and standard deviation **sigma**. You can see the data in the small graph at the top. A *t*-test of the mean appears in the middle of the screen (with **Verbose** turned off, to save space). On the right, you can see a distribution of 100 *P*-values from 100 repeated tests, each test a new sample drawn from that normal distribution.

- Look at the test in the middle of the screen and the **data** above it; in particular, look at the data values and the *P*-value from the test; make sure they seem to correspond. That is, the data probably straddle zero, and the *P*-value is probably high.
- Drag the **mu** slider to the right, say, to 1.0 and see how the *P*-value changes. (Note: the histogram of the distribution of *P*-values will not change.)
- Drag **sigma** to make it smaller; again, see how *P* changes.
- This completes our orientation; return the sliders to **mu = 0** and **sigma = 1**—a standard normal distribution, and one where the null hypothesis ($\mu = 0$) is *true*.
- Click the **Collect More Measures** button in the collection at the lower right. Fathom empties the graph and starts performing *t*-tests. Observe how the top graph updates with new samples, the test updates with each analysis, and the histogram updates with each new *P*-value. Do this repeatedly until you understand what's happening. See if you can figure out the distribution of *P*.

Probably, with only 100 points, it is not clear what that distribution is. In fact, it's *flat*, which is a very important result you may never have seen before. Now, in order to see this better, we're going to increase the number of tests we do from 100 to 500. But with the animation, it will take too long. So the next two steps are for speed; you can omit them if you have a Really Fast computer.

When you do these steps, however, you will no longer see the test itself, the data, or the gradual updating of the histogram. Slow it down again if you need to for understanding.

- *Speed step one*. Turn the graph of the original data (the one with three points) and the *t*-test into icons by dragging their corners until they're small.
- *Speed step two*. Double-click the collection **Measures from Test of data** (the box, not its name) to open its inspector. Turn animation *off*, and increase the number of measures from 100 to 500. Close the inspector.
- Now click **Collect More Measures** to get a fresh, highly-populated plot. It should look something like the illustration (and more convincingly flat).

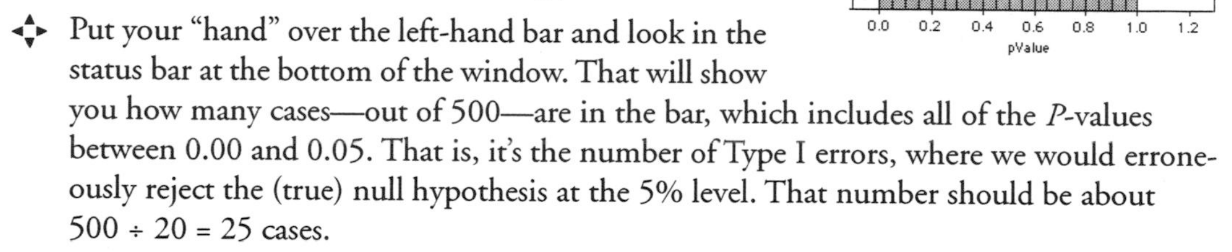

- Put your "hand" over the left-hand bar and look in the status bar at the bottom of the window. That will show you how many cases—out of 500—are in the bar, which includes all of the *P*-values between 0.00 and 0.05. That is, it's the number of Type I errors, where we would erroneously reject the (true) null hypothesis at the 5% level. That number should be about 500 ÷ 20 = 25 cases.
- Predict what will happen when you increase **mu**. (Remember what happened when you increased **mu** before we collected measures.)
- Increase **mu** to 0.2 and collect measures again. Observe how the graph changes, and how the count of rejections in the left-hand bar changes as well.
- Keep increasing **mu** and collecting measures, until you reject the null hypothesis in almost all of the tests.
- Predict first, then start increasing **sigma**, collecting measures as you go. What happens?

The Point

Two things happen, really. First, accepting or rejecting a hypothesis is not the clear-cut decision we might naïvely think. In Fathom, we can control the highly-variable truth, and test repeatedly, so we see a range of results. And look how mushy the distribution of *P*-values is! Even at $\mu = 1.5$, we reject $\mu = 0$ way less than half the time. So suppose you do a study and you get $P = 0.07$. What should you make of it? Whatever you decide, remember the wide variety of "truths" that could plausibly give you that result.

Second, statistics is at its most useful in that region between the null hypothesis and where it's obvious you should reject it. That's the transition we just explored: between only 5% of the tests being in the region $P < 0.05$ and almost all of the tests falling there. Each test has a characteristic function that describes how the probability of rejection changes as the population parameter changes: this is *power*. Ideally, your test will not reject the null hypothesis when it is true, but just stray a millimeter from that knife edge, and wham! It rejects. Alas, you can't make such a test—there are always tradeoffs—but some tests come closer than others to this ideal.

Demo 44: Power

How power—the chance that you reject the null hypothesis—changes with the population parameters

Here's one way to think about it: Power is the chance that you will get what you want.

According to the dominant scientific paradigm, our desires do not enter into it. In fact, a lot of the cautionary machinery of statistics exists in order to ensure that our results are honest and dispassionate. In many cases, a null result is worthwhile and illuminating. But let's face it: you just spent a year—or ten—of your life, or $50 million, working on something that matters to you. And it all comes down, at the end, to a *P*-value. If it's small, what you did worked. If it's not, you may not be a failure as a person, but you still feel bad.

So *before* you do this work, you should know the chance that you will get that low *P*-value—the chance that you will reject the null hypothesis. Finding that probability—the power of your test—is one of the most useful ideas in elementary statistics. But it's conceptually difficult, too.

First, it depends hugely on what the truth is. Suppose your test boils down to a null hypothesis that the mean of some quantity is zero, and you test whether you can reject that hypothesis with an α level of 0.05. If the true mean is 0.0001, the null hypothesis is false. But if the spread of the data is about 1 unit, you won't reject the null hypothesis any more than if the mean were zero. On the other hand, if the true mean is 10, you'll reject it every time.

Importantly, power also depends on sample size. The bigger the sample, the better you know the true mean—and the more likely you are to detect a difference in that mean from the null value. Finally, power depends on the α level, as we shall see. If you are willing to accept more false positives (Type I errors), you can reduce your false negatives (Type II errors), thereby increasing your power.

So power depends on the true value, on sample size, and on the significance level α. It also depends on the spread of the population. While you'll often see it reported as a single number, it's important to remember that power is a *function*.

What To Do Open the file **Power.ftm**. It will look something like this:

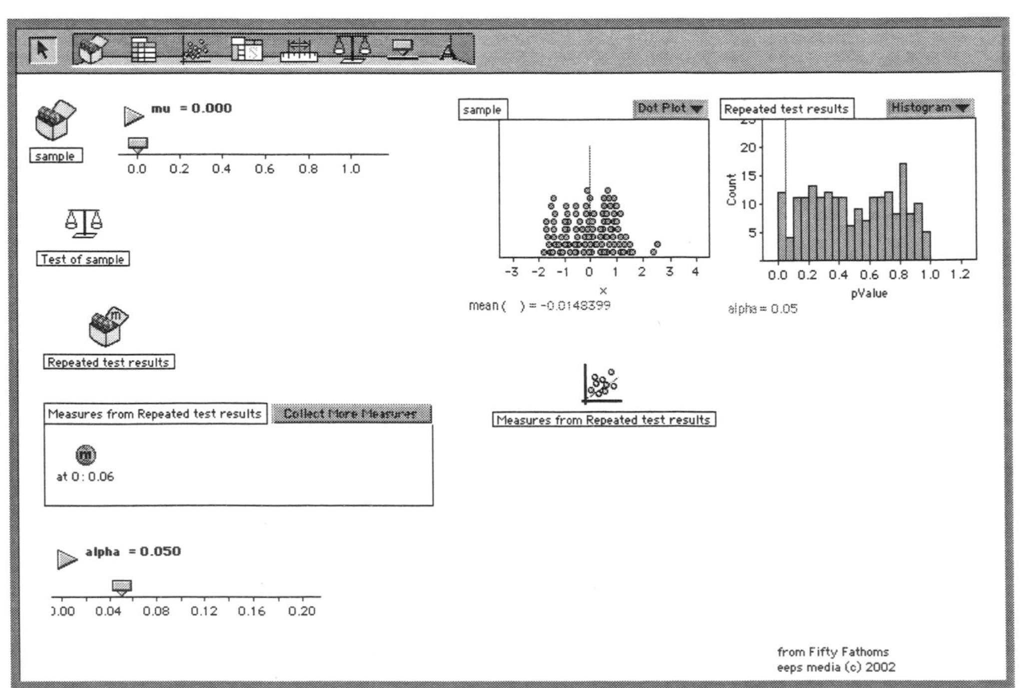

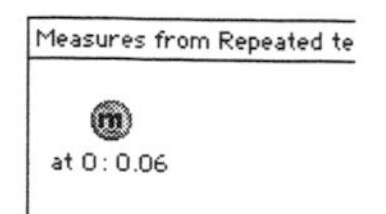

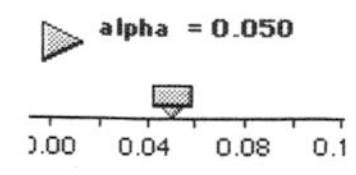

The machinery in this demo is complex, and some objects have been shrunken into icons. Be sure you have looked at “The Distribution of P-values” on page 142.

Sample holds 100 cases, graphed in the top-center dot plot. Their true population mean is **mu**, controlled by the slider. They are normally distributed with a standard deviation of 1.0. The test is a one-sample *t* test of the mean against a null hypothesis that $\mu = 0$. The *P*-values from 200 such tests are graphed in the histogram, upper right. The open, measures collection at the left side summarizes those 200 tests. In this case, 0.06 of them (at $\mu = 0$) rejected the null hypothesis at the 0.05 level; that level is controlled by the **alpha** slider just below it.

- Press the **Collect More Measures** button in the open collection. Fathom performs 200 tests, resampling every time, and updates the displays. A new summary “ball” appears.

Make sure you understand where the number on the ball comes from: it’s the number of cases in the histogram that are less than 0.05, divided by 200 (the total number of tests). Furthermore, make sure you understand that these are tests where we *wrongly* rejected the null hypothesis: Type I errors. Finally, note that these values are not far from 0.05—which is good because that’s exactly what the alpha level is supposed to be: the chance that you get a Type I error.

- Change **mu** to a small value, about 0.1. See how the dot plot changes.
- Press **Collect More Measures** again. Graphs update, and you get a new ball.
- Let’s start graphing these results: open the iconified graph by dragging its lower-right corner to fill much of the remaining space. You’ll see the proportion of rejections, **power**, plotted against **trueMu**, the population proportion.
- Repeatedly change **mu** and **Collect More Measures**, filling in the graph pretty well between **mu = 0** and **mu = 1**. Make sure at every point that you can see the correspondence between the new point and the histogram above (which you may need to rescale from time to time). You should see a graph like the one in the illustration.

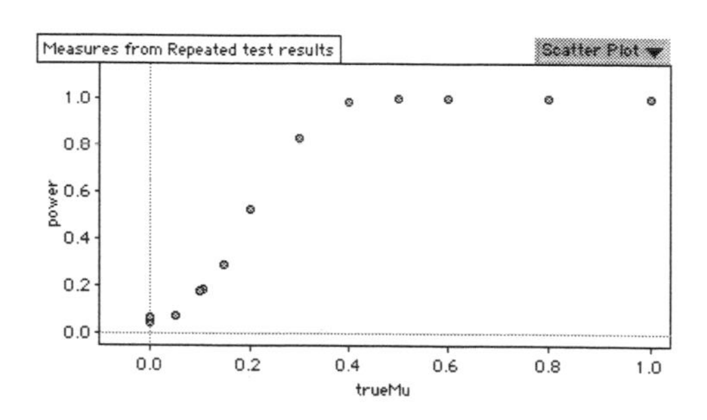

Notice that the height of each point is like the height of the left-hand bar in the histogram of *P*-values—the bar for the region below $P = 0.05$— which represents the tests where we would reject the null hypothesis.

This curve is one way to look at the power function. What does it tell you? That if you’re taking samples of 100 from this population with a standard deviation of 1.0, and your standard for detecting an effect is $P < 0.05$, your chances of rejecting that null hypothesis are good if the true population mean is over about 0.4, and lousy if it’s less than about 0.2, even though the null hypothesis would be false. In practice, many people accept powers of about 0.8 or 0.9—corresponding here to a mean of 0.3 or 0.4.

Let’s see what the results would be if we used a different alpha.

- Set the slider **alpha** to **0.2**.
- Repeat the previous actions, filling in the graph between **mu = 0** and **mu = 0.5**. You’ll see the new points appear on the graph, coded to show the different **alpha**s (here called **sigLevel**).

Here you see that we can detect a smaller difference from zero in the mean. For example, if the population mean **trueMu** is only 0.2, we would have a **power** greater than 0.8.

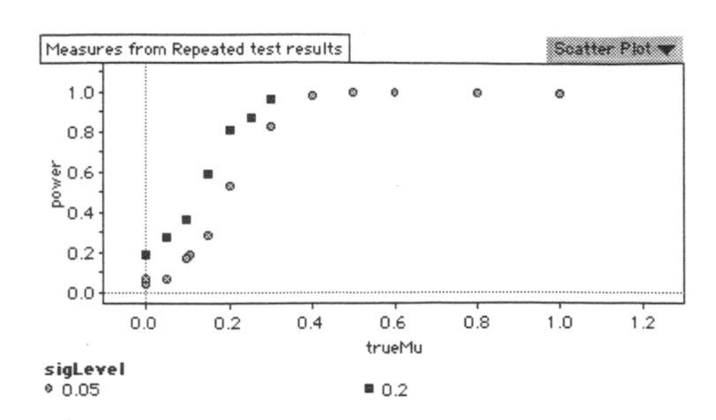

This graph illustrates two important principles:

- The intercept of the power curve is the alpha level of the test.
- If you increase the alpha level, you increase the power.

It's All Hard Stuff

1 Explain why the alpha level is the same as the intercept.

web 2 What would happen to these curves if the population standard deviation were 2.0 instead of 1.0?

web 3 The points in the graph look as if they don't quite lie on a smooth curve. Why not?

4 It looks as if the curve—the one the points don't quite lie on—has a slope of zero at the vertical axis. Is that true? (Zoom in, add points, and see what you think.) Explain why or why not.

5 Recreate this demo, but instead of using the *t* test, reject the null hypothesis if the box in a box plot of the sample data does not overlap zero (use the functions **Q1** and **Q3** to determine that formulaically). Compare that to the *t* test. Which is more powerful?

6 Recreate this demo, but with a test of difference of means. (You'll need to create a "null" sample in your source collection to compare to.) What do you find? How can you relate it to the test of mean?

7 Recreate the previous task (difference of means) except that this time, instead of using *t*, reject the null hypothesis if the boxes in the two box plots do not overlap (use the functions **Q1** and **Q3** to determine that formulaically). Compare that to the *t* test. Which is more powerful?

Demo 45: Power and Sample Size

How power—the chance that you reject the null hypothesis—changes with sample size

In the previous demo, "Power" on page 144, we saw how power—the chance that we'll get what we want (a rejection of the null hypothesis)—depends on the population parameter and on the significance level α of the test. But it also depends on sample size; that's what this demo is all about. This issue is critically important for designing experiments. You have to figure out how big a sample you will need in order to demonstrate what you want to show. If your sample is too small, the whole experiment may be useless. Put in economic terms, if you have low power, you're certain to spend money but are unlikely to get a significant result. On the other hand, samples can be expensive: too much data and you spent more than you needed to.

We begin with a file based on the one from last time…

What To Do

- Open the file **Power and Sample Size.ftm**. It looks something like this:

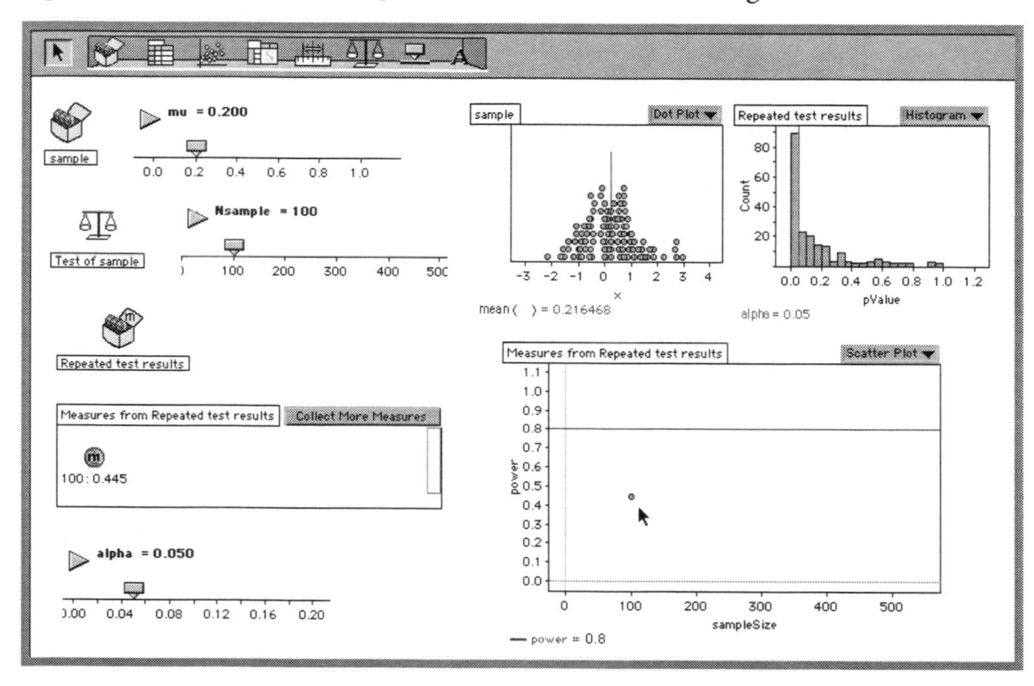

This file is like the last one except that we have a slider for the sample size, **Nsample**, currently set to 100. We have also set the population mean **mu** to 0.20. The test is still against $\mu = 0$ (which is false). So the power we see, 0.445, is the chance that we will correctly reject the null hypothesis; it corresponds to the 89 cases in the left-hand bin of the histogram (which still summarizes the *P*-values from 200 *t*-tests). Notice that the graph is now **power** as a function of **sampleSize**.

That power, 0.445, is not enough to convince us that we should go ahead with the experiment. How big a sample do we need to have an 80% chance of rejecting $\mu = 0$ if the population is at 0.2?

- Change **N** to 150. Then click **Collect More Measures** in the **Measures from Repeated test results** collection. Fathom resamples from the population, performs 200 tests on the samples of 150, and reports the *P*-values in the histogram. Another point appears on the scatter plot.

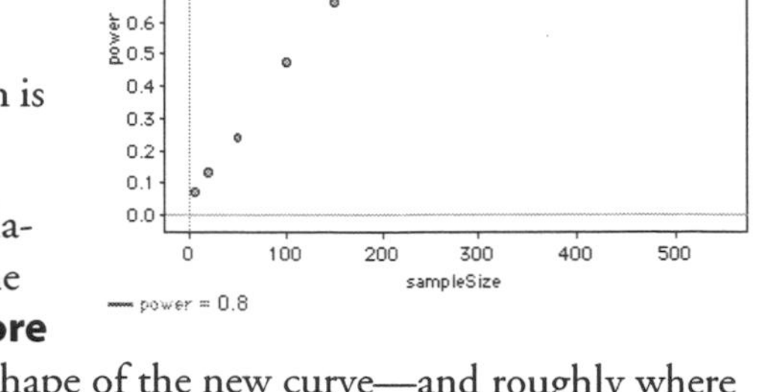

- Do the same for other values of **N** ranging from 5 to 400. You should wind up with a graph looking something like the illustration; you'll see that to get a power of 0.8 requires a sample of somewhere around 200.

Of course, that is true only if the population mean is really 0.20. What if it's different?

- Set **mu** to 0.5 (that is, what if the true population were further from zero?). Then repeat the process, changing **N** and pressing **Collect More Measures** alternately until you can see the shape of the new curve—and roughly where it reaches a power of 0.8.
- Do the same with **mu** set to 0.15.

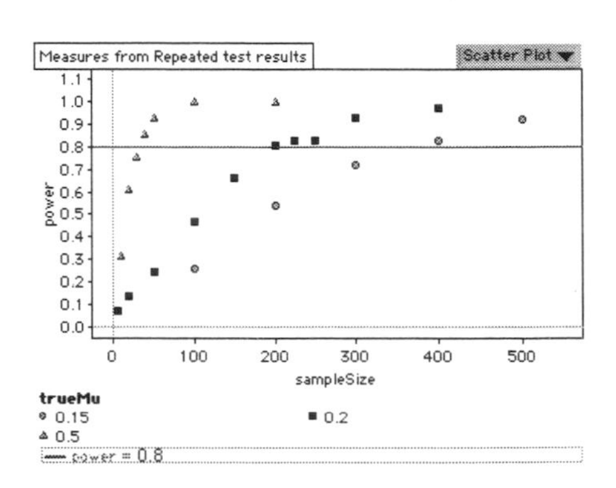

When you change **mu** and collect measures, a legend appears. When you're done, the graph will look like the one in the illustration.

What does this tell us? If the effect is a lot larger (**mu = 0.50**), we can get by with a much smaller sample of about 35 cases, and still have that 80% chance of showing that the mean is not zero at the 5% level. On the other hand, if we're wrong in the other direction by only a little bit (**mu = 0.15** instead of **0.20**), we'll need roughly twice the sample size to get that power of 0.8.

Only harder stuff, again

web

1. In this situation, it looks as if the curves do not have a slope of zero when they hit the axis (**sampleSize = 0**). Does that make sense? Try to explain it.
2. What would happen to these curves if the population standard deviation were 2.0 instead of 1.0?
3. Suppose we made a graph of the sample size you need for a power of 0.80 as a function of the population mean. What would that look like, roughly? How could you use it to plan experiments?
4. What do you think about looking for a power of 80%? What considerations would make you want it to be higher? Or lower? How high should it be before you do an experiment?
5. As at the end of the previous demo, recreate this, but instead of using the *t* test, reject the null hypothesis if the box in a box plot of the sample data does not overlap zero (use the functions **Q1** and **Q3** to determine that formulaically). Compare that to the *t* test. Which is more powerful?

Demo 46: Heteroscedasticity and its Consequences

Homoscedasticity is an assumption behind many statistical calculations. What happens when that assumption is not met?

When you calculate a least-squares line, you get the slope and intercept. Suppose you get a slope of 0.1. Is that significantly different from zero? To find out, you do a test of that slope (or estimate it with a confidence interval). That test has behind it an assumption of *homoscedasticity*—that the variance of the *y* variable is the same for any value (or infinitesimally thin swath) of *x*. Usually it's so hard to spell it[1] we fail even to check. But we should; and this demo shows us why.

First, though, why is this demo here in the section on Power? Because to understand it, you need to have delved deeply into the meaning of a statistical test. In particular, you need to know that, if the assumptions are met and the null hypothesis is true, the distribution of *P*-values for repeated tests will be uniform. We learned that in "The Distribution of P-values" on page 142. Without knowing that, it's hard to explain why homoscedasticity is important. After all, the procedure for getting the slope will give you the right answer on the average if the data are heteroscedastic. But the inferential statements you make, as we shall see, will be incorrect.

What To Do

✧ Open the file **Heteroscedasticity.ftm**. It looks something like this:

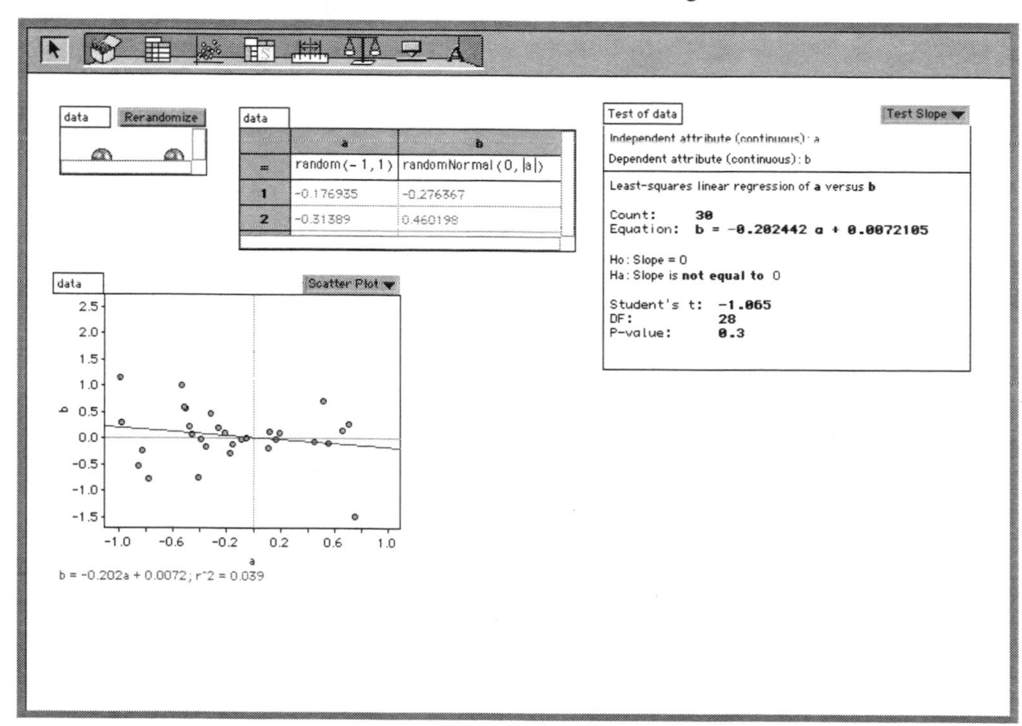

In this file, we have a collection, **data**, with two attributes, **a** and **b**, plotted against one another. Both are determined randomly. You can see their formulas in the "=" row of the case table. The attribute **a** is uniform in the range (–1, 1); **b** is normally distributed with a mean of zero—whatever the value of **a**—but its standard deviation gets larger the farther it is from the axis. That is, its true distribution looks like butterfly wings or a bow tie, and the true slope of **b** on **a** is zero—it must be because its mean value is zero whatever the value of **a**. The test,

1. Some people do so with a "k," but Random House disagrees, and even lists the pronunciation with a soft (or silent) "c": ho-mo-se-das-*ti*-si-ty. By the way, some luminaries prefer *homogeneity*, but *homoscedasticity* is a cool word, even if it is weird. Both are correct, but homoscedastic (having equal variance) is more specific than homogeneous (having a common property throughout).

upper right, is testing whether the slope is not equal to zero; you can see the *P*-value of 0.3 in the illustration—your file will be different. That value suggests that we should *not* reject the null hypothesis (which is good, because it's true).

- Press **Rerandomize** (in the data collection) repeatedly. The graph and test will change.
- Look informally at the slopes. For example, count how many slopes are positive and how many are negative in ten tries. You should be able to detect no obvious bias.
- Do the same for *P*-values. You will probably judge that they are spread pretty well over the range [0, 1].

But there *is* nonuniformity; we just need a way to detect it.

- Choose **Show Hidden Objects** from the **Display** menu. An empty collection, **Measures from Test of data**, and a graph, appear.
- Press the **Collect More Measures** button on the collection. Fathom churns, and 20 points appear in the graph. They'll be way down at the bottom. They are *P*-values from the 20 tests Fathom performed on 20 new samples.

Since the null hypothesis is true, this distribution of *P*-values should be uniform. But is it?

- Do it again, a few times, watching the distribution begin to grow.
- Finally, double-click that measures collection to open its inspector. It should open to the **Collect Measures** panel; if it does not, click on the tab to take you there. Change the panel to turn off animation and to collect 200 points at a time. Close the inspector.

For extra speed, iconify the test object in the upper right, or even the scatter plot. (Do so by dragging a corner until they're small.) Then Fathom does not have to redraw them.

- Again press **Collect More Measures** a few times, and build up the distribution so you see a clear nonuniformity. Your graph will eventually look something like the illustration.

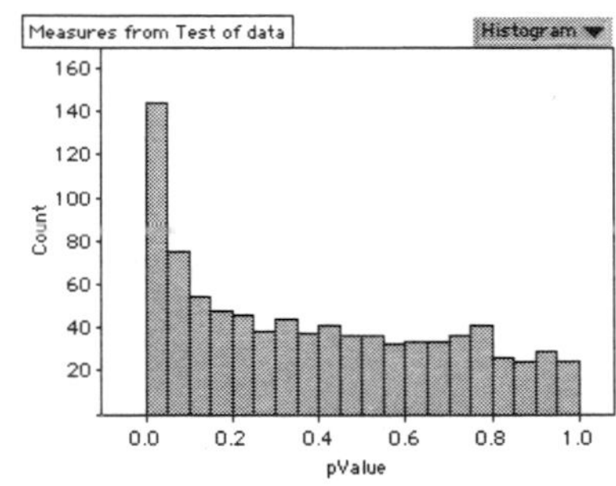

That is, it seems as if the test gives a low *P*-value, rejecting the null hypothesis, more often than it would if the distribution were uniform. How much more often? Let's see:

- Select the histogram by clicking on it once. Then choose **Scale>Relative Frequency** from the **Graph** menu. In our graph, it showed that the left-hand bar—the one showing $P < 0.05$—had a relative frequency of 0.16.

What does this mean? Suppose we said we would reject the null hypothesis with a *P*-value of less than 0.05. Even though the null hypothesis is true—**b** does not depend on **a**, so the slope relating them is in fact zero—we would reject it 16% of the time. That is, *we would get 3 times as many Type I errors as we should.* That means that the whole logic of the hypothesis test is wrong; if you get a low *P*, you shouldn't be as confident that the slope is not zero.

This can tilt the other way, too:

- Open the file **Heteroscedasticity 2.ftm**. It will look a lot like the first one except that the **Measures from test of data** collection is already showing.

In this file, the distribution of **b**, instead of looking like a butterfly, looks like a diamond: it has a wider variance in the middle, and narrower at the ends. See the formulas to understand how it's done.

- Press **Rerandomize** on the **data** collection a few times to see the scatter plot and its least-squares line update, as before. Note the values in the test, too.
- Press **Collect More Measures** in the **Measures from test of data** collection. Twenty more data points join the collection, and the bars grow in the histogram. Do so a few times to watch the distribution of *P*-values begin to develop.
- As before, open the inspector, turn off animation, and increase the number of measures to 200. Press **Collect More Measures** to build up the distribution.

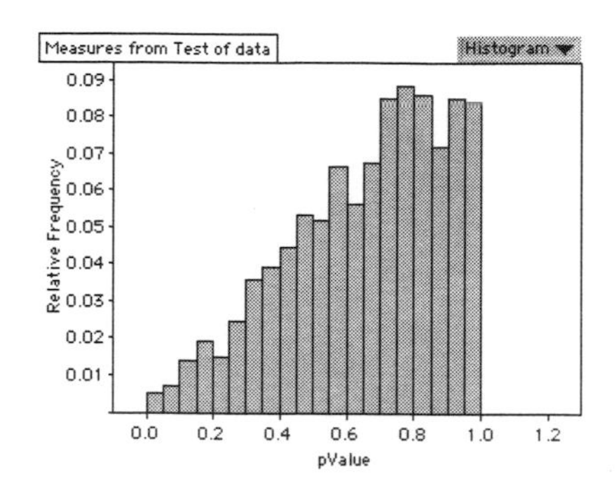

- Select the histogram and choose **Scale>Relative Frequency** from the **Graph** menu. The graph will look something like the illustration.

This time, the distribution of *P* is tilted the other way. That is, a Type I error is extremely rare. That also means that, if the true slope were *not* zero, it would be hard to reject the null hypothesis: that is, the power of this test is smaller than it would be if the data were homoscedastic.

More Hard Stuff

web

1. Explain qualitatively why the slopes in the "butterfly" distribution would give you too many significant results, and why the slopes in the "diamond" distribution would give you too few.
2. Suppose we wanted to correct the test for the heteroscedasticity. In this latest, "diamond" distribution case, what value of *P* had about 5% of the cases below it? (It looks like about 0.2; see if you agree.) So then could we say that we get a significant result (at the 0.05 level) if $P<0.2$?
3. Do the same for the "butterfly" case. What is the critical value for *P* if you want a "true" alpha of 0.05?
4. In the "diamond" case, alter the simulation so that **b** really does have a slope. That is, make a slider **m**, and make **b**'s formula **m*a + randomNormal(0, 1−|a|)**. How does the distribution of *P*-values change as **m** increases? Explain what you find.
5. We commented that the power of the test in the diamond case was smaller than it would have been if the distribution were homoscedastic. Does that mean that heteroscedasticity made the test in the butterfly case *more* powerful? Explain this; that is, if you think it's more powerful, wouldn't we always want to encourage butterfly heteroscedasticity?—that is, if we could pronounce it…
6. One of our favorite luminaries (Chris Olsen), on reviewing this demo, suggested that, while the diamond and butterfly examples were good, the more common way you get heteroscedasticity in practice is when the width of the error distribution is in proportion to the measure. That is, the distribution looks more like a trumpet. Predict first, and then simulate: what difference would that kind of distribution make in the distribution of *P*? (Simulation hint: Change the formula for **a** to be positive only, then look at the formula for **b** to see if you need to change it as well.)

Distributions

In the course of these demos, we have come across several distributions—the uniform distribution, the normal, the binomial, and Student's *t*. Of course, there are a slew of other distributions we have not seen; in this thin book we can't cover everything. But let's take a look at a few of these "other" distributions and get a whiff of what they might be useful for. Most importantly, we will see that many interesting things are not normally distributed, or even close.

This section includes:

"Wait Time and the Geometric Distribution" on page 154. How many rolls does it take to get a six? How many deals will I have to wait through until I get a blackjack? With a fair die and an infinite deck, these are distributed *geometrically.* This distribution appears in a wide variety of situations, and has wonderful properties.

"The Exponential and Poisson Distributions" on page 157. The continuous analog to the geometrical distribution is the *exponential*: how long will I have to wait for the next phone call, raindrop, or nuclear decay? (That's also the distribution of times *between* events.) And then, if we ask how many events will I have in the next ten minutes, those numbers will be distributed according to the *Poisson*.

**"Sampling Without Replacement
and the Hypergeometric Distribution" on page 160.** You know how we always draw cards from infinite decks? Suppose it's not infinite. Or suppose your sample size is a significant fraction of your population? Then you need the *hypergeometric* distribution. Without it, card counting doesn't work.

"The Bizarre Cauchy Distribution" on page 163. For our last demo, we'll look at something really pathological: a distribution that has no standard deviation. The samples have one, but the distribution does not. How can that be?

web — Remember: If you see web in the side margin, that means that there's a solution on the web at the time of publication. See **http://www.eeps.com**.

Demo 47: Wait Time and the Geometric Distribution

The distribution of times until something happens • How this is the geometric distribution

Urns are used for three things in our society: holding tea (rarely); holding the ashes of the departed; and holding colored balls from the probability problems in mathematics textbooks. In this demo, we will draw balls from an urn until we get a red one. Every time we draw out a ball, we will put it back before drawing the next one.

Here we are interested in the distribution of *wait times*—how long it takes (measured in numbers of balls) until we get the first red. We'll find it follows what's called the *geometric distribution*, so called because the counts at each successive draw follow a geometric sequence. It's also interesting to find the mean of this distribution, as that is the expected number of draws (in the "expected value" sense of that word).

What To Do

- Open the file **Wait Time.ftm**. It should look like this:

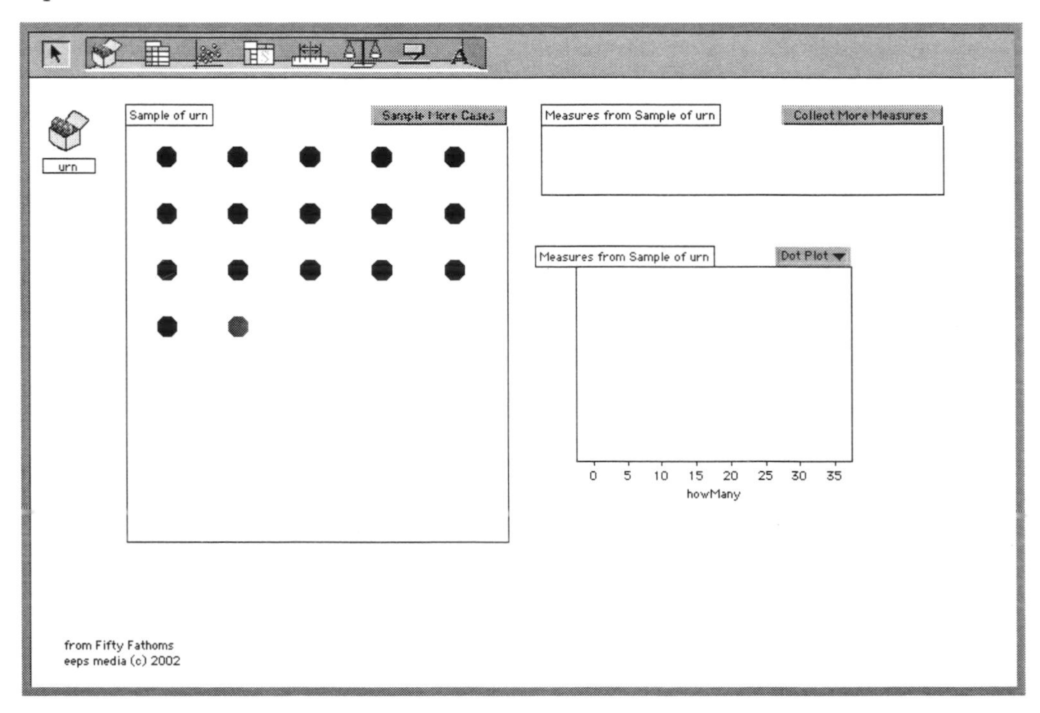

Here you can see the **urn**—the collection in the upper left. Open it up if you like by dragging its lower-right corner. You'll see that it contains ten balls, only one of which is red. But then close it to save space.

The next collection is the **Sample of urn** collection; it contains the records of balls you draw—with replacement—from **urn**. You draw until you get a red one. The other two objects we will get to shortly.

- Press the **Sample More Cases** button in the **Sample of urn** collection. Fathom will draw new balls from urn until you get a red one. Note how many draws it takes.
- Do this a few times, until you're sure you understand what's going on, and you know what kinds of numbers of draws you see until there's a red ball.
- It would be great to get Fathom to record these automatically. Press the **Collect More Measures** button in the collection (empty box) in the upper right. Fathom's machinery

goes to work, and you see measures (the number of balls, called **howMany**) appear in the collection and in the graph below.

- Every time you press that button, you get five more measures. Keep doing this until you start to get an idea of the shape of the distribution.
- Let's figure out the mean. Click once on the graph to select it, then choose **Plot Value...** from the **Graph** menu. The formula editor opens.
- Enter **mean( )**, then press **OK** to close the editor. The mean will appear, and the graph will look something like the illustration.

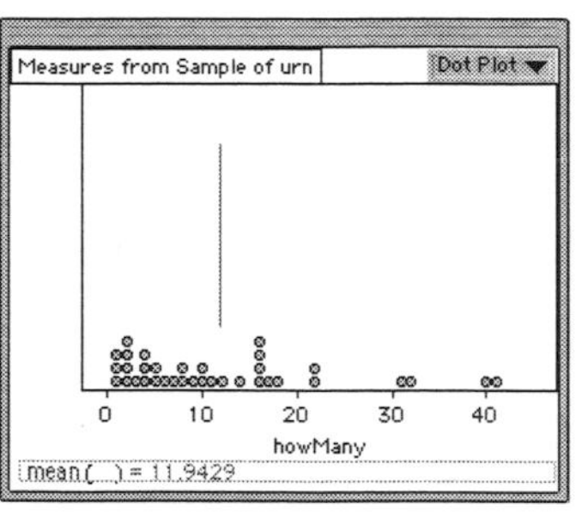

By this time, you may be getting impatient with the flying gold balls. Let's get rid of them to speed things up:

- *Speed step 1*: Double-click the sample collection (the big one with the colored circles) to open its inspector.
- *Speed step 2*: Un-check the **Animation** box in the **Sample** panel and close the inspector.
- *Speed step 3*: Double-click the measures collection (the one with gold balls labeled **n = 3**, etc.) to open its inspector.
- *Speed step 4*. Un-check **Animation** here as well.

Now that we can go faster, let's add a lot of measures to that graph:

- In that **Collect Measures** panel, change the number of measures collected from 5 to 50. Then close the inspector.
- Press the **Collect More Measures** button to add 50 cases to the dot plot. (You can make the measures-collection go even faster if you make the **Sample of urn** collection so small that it turns into an icon. That way Fathom doesn't have to redraw it.)
- Keep doing this until the mean seems to converge to a sensible value (the theoretical value is 10.0). If you get too many points in the dot plot, change to a histogram (as shown at right).

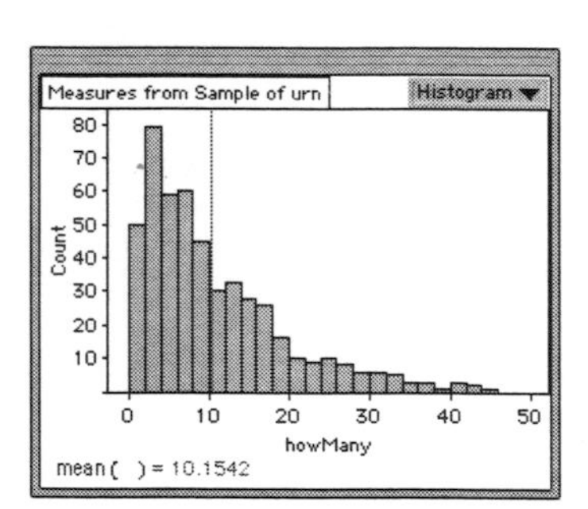

Note: If you make a histogram, it will probably have a bin width of 2, like the one in the illustration. This is a problem only in the first bin, which contains both 0 and 1. Since 0 is impossible (you can't draw a red ball on the zeroth try), that first bin has many fewer cases than the neighboring bin. This misrepresents the ever-decreasing distribution.

Information about this graph:
Histogram: Bin width: **2.0000** starting at: **1.0000**
The **howMany** axis is horizontal from **−2.5000** to **52.500**
The **Density** axis is vertical from 0 to **0.11624**

If this bothers you, force the binning to start at one. Double-click the graph; a *ControlText* block appears. Change the text (as in the illustration) so the bins are "starting at 1.0" and press **tab** or **enter**. Then get rid of the Control-Text if you like: select it and choose **Delete Control Text** from the **Edit** menu.

Questions

1. Did the mean converge to 10.0—or at least close to it?
2. How would you describe the shape of the distribution?

3 Why would each number be less likely than the one before?

Extensions

Let's plot the theoretical distribution.

- Make sure your graph is a histogram. Then, with the graph selected, choose **Scale>Density** from the **Graph** menu. The vertical scale will change.
- Choose **Plot Function...** from the **Graph** menu. The formula editor opens.
- Enter **geometricProbability(floor(x), 0.10, 1, 1)**. Press **OK** to leave the formula editor. The function appears as in the illustration.

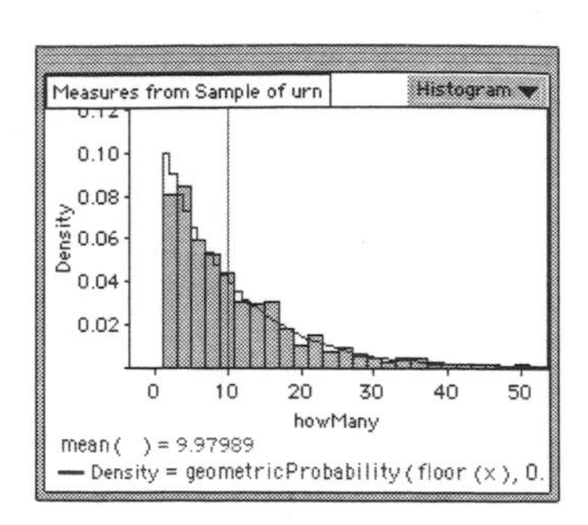

Of course there are other interesting things to do. One of the best is to change the probability of getting a red ball. Do that by altering the original **urn** collection. For example, to get rid of a black ball, just select it in **urn** and then choose **Delete Case** from the **Edit** menu. To make a new ball, make a new case and give its attribute **tag** a value of **1**.

Then, before you **Collect Measures** to see the distribution of wait times, delete all of the cases in the **Measures from Sample of Urn** collection (they're from a different population). To do that, select them (select the graph and choose **Select All Cases** from the **Edit** menu); then **Delete Cases** in the **Edit** menu.

Harder Stuff

4 Make a conjecture about the mean of any geometric distribution.

web 5 Prove that conjecture. (It helps to know the sum of an infinite geometric series but you can do without if you know a trick.)

web 6 Is the *median* of a geometric distribution larger, smaller, or the same as the mean? Use Fathom to demonstrate that what you say is correct. Also, explain why it is correct based on the shape of the distribution.

web 7 What is the *mode* of a geometric distribution? Explain.

Theory Corner

We derive the theoretical expected value (i.e., the mean) for the number of trials in "The Geometric Distribution: Proof That the Mean is (1/p)" on page 180.

Demo 48: The Exponential and Poisson Distributions

The continuous analog to the geometric distribution • How many events happen in a given period: a Poisson distribution

Suppose phone calls come in at random times, on the average once per minute. How many seconds will it be until the first one? On the average, 60—but what is the distribution?

Based on our experience with the geometric distribution, we could say that the probability of a call during the first second is 1/60. Using that, we could simulate it one second at a time. But what if we could time the ring to the nearest 1/10 of a second? Then we would use a probability of 1/600, and use 1/10 second as our time step. Can you see where this is going? There must be a way to make this discrete phenomenon—repeated trials with a constant probability—continuous. And there is; instead of the geometric distribution, we'll use its continuous cousin, the exponential.

What To Do

❖ Open the file **Exponential and Poisson.ftm**. It will look something like this:

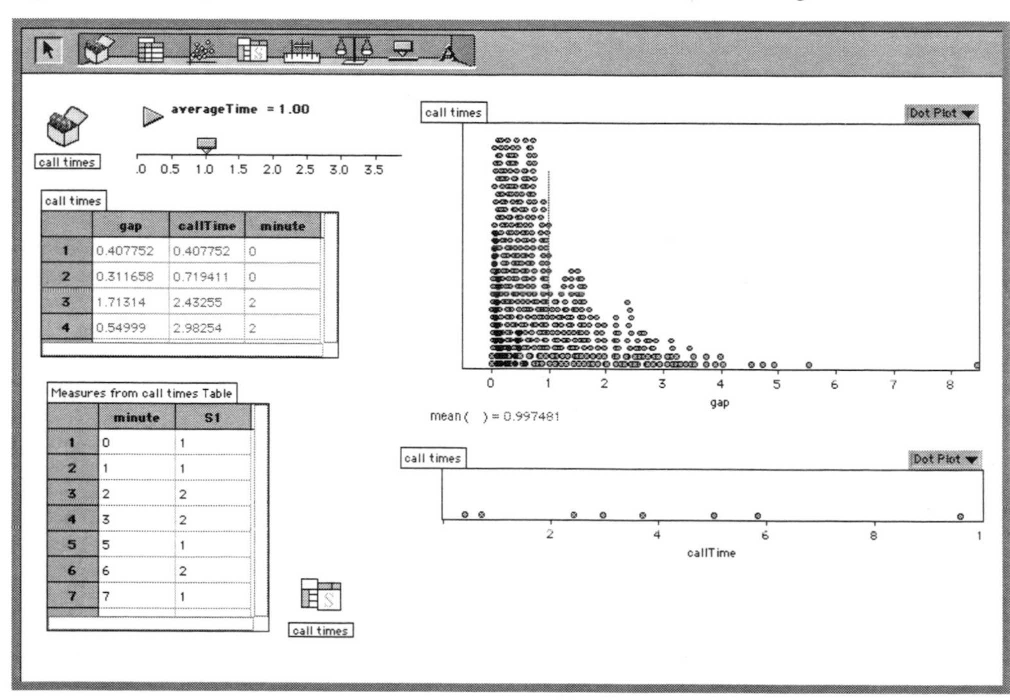

Since the calls are independent, the distribution of times *between* calls is the same as the distribution of times of the *first* call: exponential.

Let's concentrate on the upper half of the window. The slider **averageTime** controls the average time between calls. The **call times** table below it shows the first 4 of 500 calls. Attributes are the **gap** (the time between calls); the **callTime** in minutes, which is the time in decimal minutes that this call came in; and **minute**, which is the integer part of **callTime**, so it tells us *which minute* that call came in. The big graph shows the gaps, whose distribution decreases exponentially—whence the name of the distribution. Notice that, while the average time between calls is 1, most of the gaps are shorter. Below the graph is another showing the times of the first few minutes' calls.

One last word of warning before we begin: Fathom is doing a lot of computation under the hood here, so be patient. It can take *several seconds* (it seems like years) for the display to update, even on a fast machine. Don't think prematurely that you've crashed. Also: you will save yourself a lot of grief in this simulation if, when you want to change the slider value **averageTime**, you do so by editing the value, *not* dragging the pointer with the hand.

- Click on the **call times** collection in the upper-left corner to select it.
- Choose **Rerandomize** from the **Analyze** menu a few times (unless you have a fast computer, waiting for the display to update) to see how the distribution looks.
- Change the distribution to a histogram by choosing **Histogram** from the popup menu in the upper-right corner of the graph. Note: you can always rescale the axes by re-choosing **Histogram**.

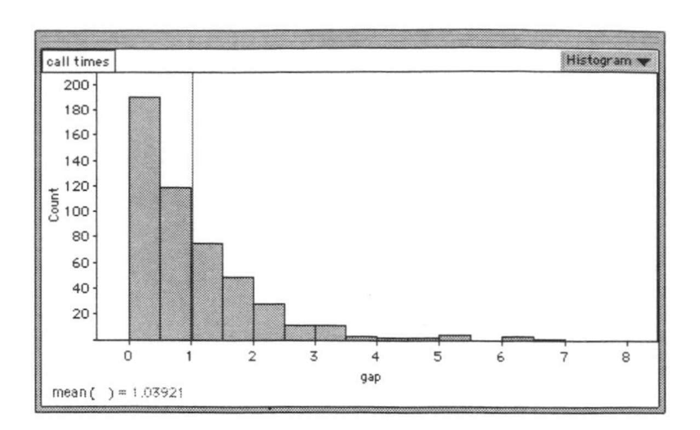

Now let's look at the machinery in the lower left. There is an iconified summary table and case table derived from it (**measures from call times Table**), which shows what we want to see. It shows how many calls came in during each minute. In the picture on the previous page, you can see that one call came in during the first minute (minute zero), and one during the second, and so forth. Note that no calls came in during minute 4. You can see that gap in the short, wide **callTime** graph and, if you wish, in the upper case table (you will have to scroll down). By this point your particular display will be different from the illustration; find analogous places in your current data.

- Change the value of **averageTime** to **2.0**. Remember to edit the value and then press **tab** or **enter**, given the slow response. Note how the distribution changes. Also see how the call times in the lower graph get farther apart.
- Now change the value of **averageTime** to **0.5**. Again, see how the distribution changes—and how the calls are smushed together. Try other values as you see fit.
- **Rerandomize** a few more times. (The upper collection must be selected.) Watch the lower graph and the table below to see how the gaps map onto the positions of the calls.

Let's move on. Now we're wondering, if calls come in on the average 5 per minute, how often will I get 10 calls in the same minute?

- Set **averageTime** to 0.20. Since we have 500 calls, that will be about 100 minutes.
- Now—for screen space—we have to dispose of that lower graph. Iconify it and move it to the side. If you have a full version of Fathom, you can delete it: select the graph by clicking once, then choose **Delete Graph** from the **Edit** menu.
- Choose **Show Hidden Objects** from the **Display** menu. A new graph appears, looking (after rescaling) something like the one in the illustration.

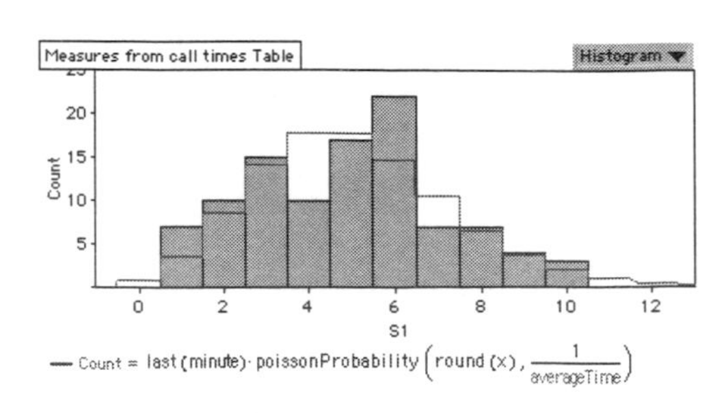

This graph shows something hard to grasp: the distribution of the numbers of calls that came in during the minutes we were studying. That is, during about 17 of the minutes, exactly 5 calls came in. During 3 of the minutes, 10 calls came in. There may even have been a minute or two where zero calls came in, but since those minutes never appeared in our data, Fathom cannot display them.

We have plotted the theoretical distribution for this situation—a *Poisson* distribution—which also gives you an idea of how many minutes had zero calls.

- Change **averageTime** to various values such as 0.5, 1.0, and 2.0. You may need to rescale both histograms by choosing **Histogram** from their respective popup menus.

Questions

1 Where did the 0.20 come from when we were simulating five calls per minute?

web 2 In that situation, after what fraction of the calls (out of 500 calls) was there a gap—the time between calls—of six seconds or less? (Six seconds is 0.1 minutes.)

web 3 If we average one call per minute (so we'll be simulating about 500 minutes), in about how many minutes do you expect to get zero calls? Remember, you'll have to think theoretically—or be sneaky—as Fathom will not display the data.

4 Same question at one call every two minutes.

Harder Stuff

5 Explain why the theoretical distribution for this situation—the Poisson distribution—has to have a tail off to the right.

6 Study the Fathom simulation and explain why it's so much faster when **averageTime** is small than when it's large.

Demo 49: Sampling Without Replacement and the Hypergeometric Distribution

How distributions change when the sample is large compared to the population

Most statistical tests and classroom problems that have to do with sampling assume that you are sampling with replacement. In real life, however, you often sample *without* replacement. For example, if you're taking a survey, you probably won't survey the same person twice. Usually, this doesn't matter a lot. If the population you're sampling from is much larger than the sample, the sample will have about the same characteristics whichever way you sample. But there are some situations where it does matter: where the population is small (e.g., a classroom), or the sample is a large fraction of the population it is drawn from.

One such setting is when you draw cards from a deck and don't put them back. We'll use a standard 52-card deck in this demo, simplified so that the cards are only **red** or **black**—26 of each. We'll draw samples of various sizes, counting how many **red**s there are. Then (of course) we'll build up sampling distributions of those numbers.

What To Do

- Open the file **Hypergeometric.ftm**. It will look something like this:

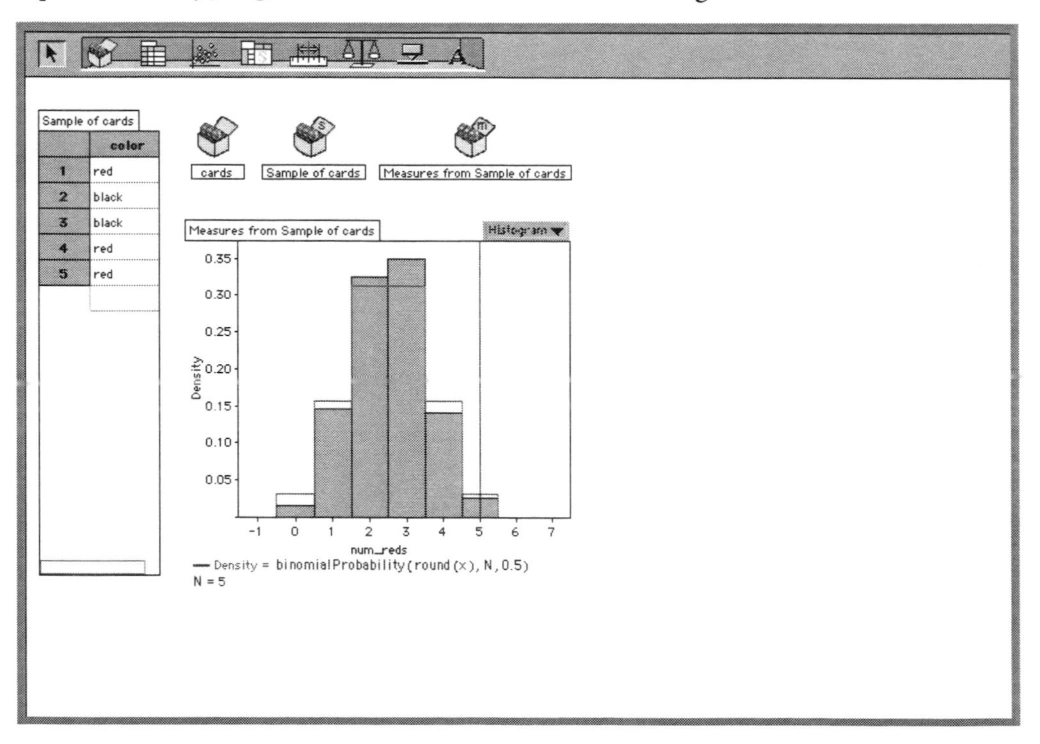

Here you can see the "source" collection, **cards**; **Sample of cards** is a sample of five cards from that collection—you can see the values in the table at the left. **Measures from Sample of cards** is a collection of 200 measures—the number of **red**s in the sample of 5, called **num_reds**.

You can see the binomial probability function plotted on the histogram of those measures, as well as a vertical blue line at the sample size—in this case, 5. That is the maximum possible value for **num_reds**; if you draw five cards, you can't get six red ones. The histogram hangs over only because the bins are fat.

- Let's get a new set of measures. Click once on the **Measures from Sample of cards** collection to select it. Then choose **Collect More Measures** from the **Analyze** menu (or press its shortcut, **clover-Y** on the Mac or **control-Y** in Windows). After a moment, the graph updates with new values.
- Repeat as necessary so that you're sure you understand what's going on.

Note: unlike in some earlier demos, we have omitted the part where you just redraw samples a few times to understand that part of this three-layer process. If you need to see it, make a graph, graph the data from one sample, and do some resampling. You might even turn on animation. We would have set that up, but we'll need the screen space, as you will see later.

You should see that the binomial function matches the histogram pretty well, which is what we would naïvely expect. But we will soon become hypergeometric cynics. Onward!

- Let's increase the sample size. Double-click the sample collection (the middle one) to open its inspector. It will resemble the one in the illustration. Note how the **with replacement** box is unchecked—for one of the only times in this book.
- Change the number sampled from 5 to 10, as shown. Click **Sample More Cases** and see how the table updates to show ten values.
- Close the inspector or move it aside so you can see the graph.

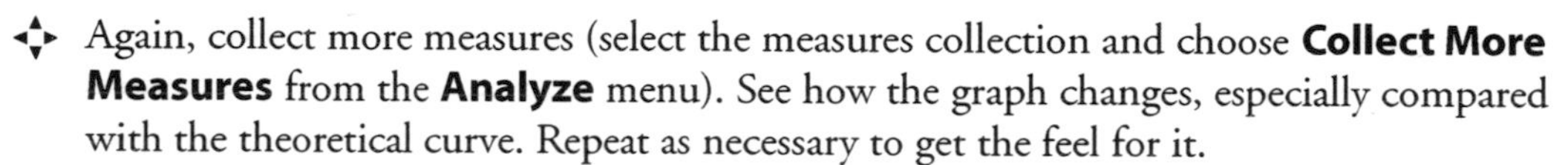

- Again, collect more measures (select the measures collection and choose **Collect More Measures** from the **Analyze** menu). See how the graph changes, especially compared with the theoretical curve. Repeat as necessary to get the feel for it.
- Do the same for sample sizes of 20 and 30.

By the time you get to $n = 30$, the function should not fit the data very well.

Questions

1. When we had a sample size of 5, there were two bars with the same theoretical value. Now there's only one. What made the difference?
2. (web) At $n = 30$, how would you characterize the difference between the data from the samples and the theoretical curve?
3. Why does that difference occur? Why does drawing without replacement change the distribution of reds you would get?
4. (web) If we sampled 52 cards instead of 30, what would the distribution look like?

Onward!

- Close the inspector if it is open (for screen space).
- Choose **Show Hidden Objects** from the **Display** menu. A new graph appears in the lower right. It has a *hypergeometric* curve on it.
- Repeat as needed, even changing sample size, to convince yourself that this distribution fits the data better than the regular binomial curve.
- If you wish, return to a sample size of 5, and see that the two distributions are nearly identical in this case.

Thus this situation produces a different distribution, more bunched together than a regular binomial distribution. If you look at the arguments of the function that is plotted in the new graph, you can see that Fathom has to know the population size (52 cards) and the number of "successes" (26 reds) to calculate the hypergeometric probability function.

Extension

- Develop a rule of thumb for when you should use the hypergeometric distribution instead of the plain old binomial. That is, when we did a sample size of 5, it looked OK, but 30 looked much different. For this, it would help to put the two functions on the same graph. You might also want to change the proportions of successes and the size of the population.

Harder Stuff

web

5 Make a Fathom document to simulate card counting in this scenario:

> Suppose you shuffle a deck and deal out 21 cards, face up. Then you bet $1.00 (at even odds) on whether the next card is red or black. The question is, if you play 100 hands, using the obvious strategy, how much do you win on average?

6 Explain clearly why this is different from playing the same game, but with flipping a coin.

7 The same as the first task, but with four decks in play. This is why they use multiple decks at casinos.

8 Rain Man. Start with a complete deck. (You could use the file **DeckOfCards.ftm** that comes with the full version of Fathom as a starting point). This time, you deal out 31 cards, then you can bet $1 or $2 on a hand of blackjack against a dealer. (This demands complicated formulas.) No hitting, but if you get a blackjack you win outright, and you lose a push. Can you win in the long run by counting face cards?

Demo 50: The Bizarre Cauchy Distribution

The Cauchy distribution • The meaning of mean and standard deviation; how it's possible for a distribution to have neither

Suppose you stand close to a very long wall, and spin around. To make matters worse, you have a pea shooter, and at random times—while you're spinning—you shoot peas. Half the time, of course, the peas go away from the wall. Of the peas that hit, many will hit near the point on the wall closest to you. But some—at angles close to ±90°—will hit the wall very far away.

The positions of the peas that hit the wall form a *Cauchy* distribution. It has a hump in the middle just like some very well-behaved distributions we know, but it has a really interesting catch: it *has no standard deviation.*

What in the world can that mean? Of course it has a standard deviation! All you have to do is take the points in the sample and compute it. True, true. But I wasn't talking about the sample; I was talking about the *population.* Let's see what I mean…

What To Do

- Open the file **Cauchy.ftm**. It will look something like this:

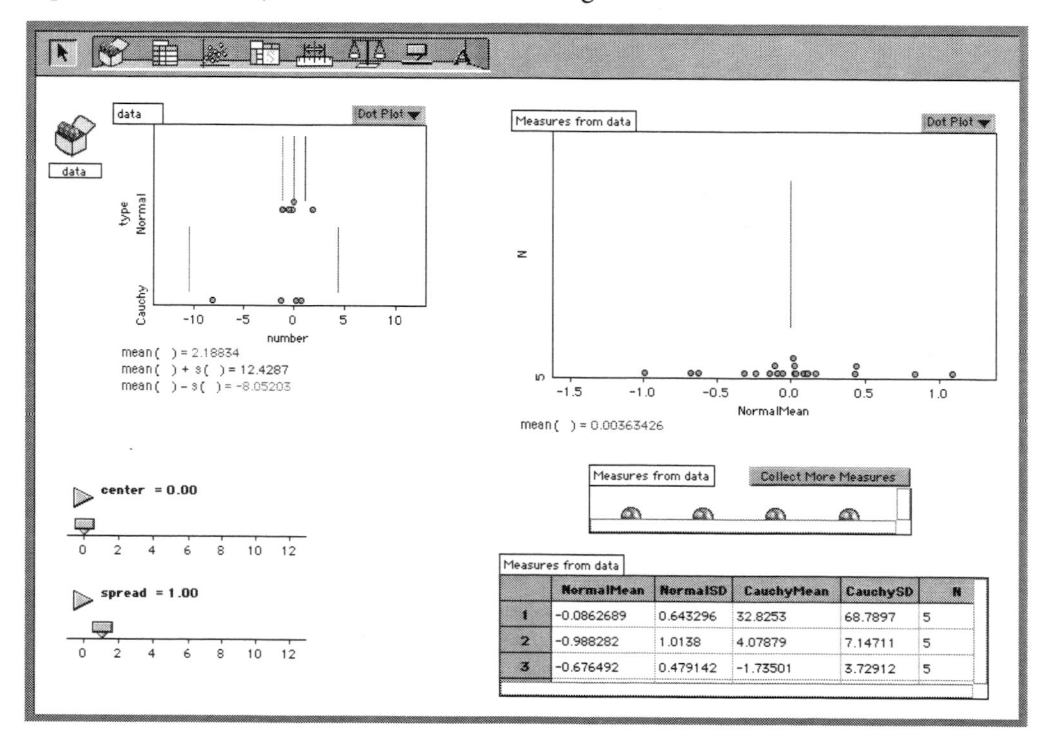

In the upper left, you see a collection and graph that shows a sample of five points drawn from a Cauchy distribution, and five more drawn from a Normal distribution for comparison. You control their centers and spreads with the sliders below (though we will not have to here).

On the right, you see a measures collection (**Measures from data**) in the middle, with a case table below it and a graph above. The graph shows the mean of the normals (the attribute **NormalMean**) from the collection on the left, from 20 separate resamplings.

- Click once on the upper-left (**data**) collection to select it. Then choose **Rerandomize** from the **Analyze** menu (or use its shortcut: **clover-Y** on the Mac or **control-Y** in Windows.) The numbers in the graph will change, as well as the lines that show the mean and ±one standard deviation of the *sample*.

- Repeat several times. Note how the Cauchy values jump around more.
- Let's add more cases. With the **data** collection selected, choose **New Cases...** from the **Data** menu; add 10 more cases and press **OK** to close the box.

Note: The way we have this set up, adding 10 cases adds 5 Normals and 5 Cauchys.

- Again, **Rerandomize** to see how the distributions change.
- On the right-hand side, press the **Collect More Measures** button in the **Measures from data** collection. Fathom rerandomizes the data collection 20 more times and adds the means and standard deviations to that right-hand collection; you'll see two distributions now: **NormalMean** for **N = 5**, and the same for **N = 10**.
- On the right-hand side, drag other attributes (**NormalSD**, **CauchyMean**, and **CauchySD**) from the case table at the bottom to the horizontal axis of the right-hand graph, replacing **NormalMean**. Observe the characteristics of each graph. The graph for **CauchySD** will look something like the illustration.

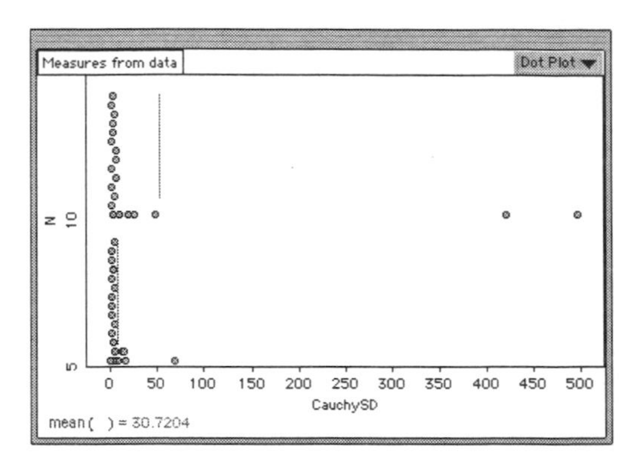

You can already see that there's something seriously weird about the Cauchy data. In the illustration, for example, while most of the resamplings have a small standard deviation, two have gigantic spreads when **N = 10**. Your statistical instincts should be saying, "we should increase the sample size; that will make these wild swings even out." Ordinarily, you would be right. But this is the Cauchy distribution...

- Alternate now between adding cases on the left and collecting measures on the right. (Remember that on the left, the number of cases in the collection is the number of Cauchys *plus* the number of Normals.) So add 20 cases (for a total of 40, or 20 of each type); then 60 (to make 100, or 50 of each type); then 100 (for 200, or 100 of each type).
- Again, drag different attributes to the horizontal axis of the right-hand graph and compare. Some of the Cauchy data will be so wild that you'll need to rescale your axes by hand to see the trends.

The Point

You should see that, as sample size increases, the means (the red lines) of **NormalMean** nicely converge to zero and that **NormalSD** nicely converges to one. But the **CauchyMean**s swing more wildly as **N** increases, and **CauchySD** *systematically increases with sample size.*

Because of this, official mathematical statisticians admit that the Cauchy—as an abstract population distribution, a probability density—*actually has no mean nor standard deviation,* because the various sums or integrals you would use to compute them do not converge. For this reason, its parameters are actually **mode** and **spread** in Fathom and in many statistics books.

Extension

Medians are supposed to tame distributions with outliers, right? Modify the document to collect medians and IQRs as well as means and SDs. You'll need to add new measures to the **data** collection, delete the cases in the measures collection, and then re-collect the measures for different sample sizes.

How Did They Do That?

Suppose you wanted to make your own demos. You might be curious about some of the fancy things we did in the files that come with this book. We used features of Fathom even experienced users might not be aware of, and these strategies should not be kept a secret. We'll treat you as an expert here, moving pretty quickly. Also, instead of telling you which menu to go to in the following instructions, we'll just allude to the context menus, which you get to using the right mouse button in Windows or **control**-click on the Mac.

Using the Collection Itself as a Graph

See "The Mean is Least Squares, Too" on page 30,
"Building the Binomial Distribution" on page 56,
"Capturing with Confidence Intervals" on page 103,
"Capturing the Mean with Confidence Intervals" on page 118,
"Fair and Unfair Dice" on page 122, and
"Wait Time and the Geometric Distribution" on page 154.

By default, the collection—the box of gold balls—represents each case as a gold ball. This is useful just as it is for its animations of how sampling works, for example. But it becomes more useful when you change those gold balls into something else.

The key is the **Display** panel in the inspector. Let's open one of the files that came with this book and look under the hood:

- Open the file **Building Binomial.ftm**. The upper-left collection looks like this:

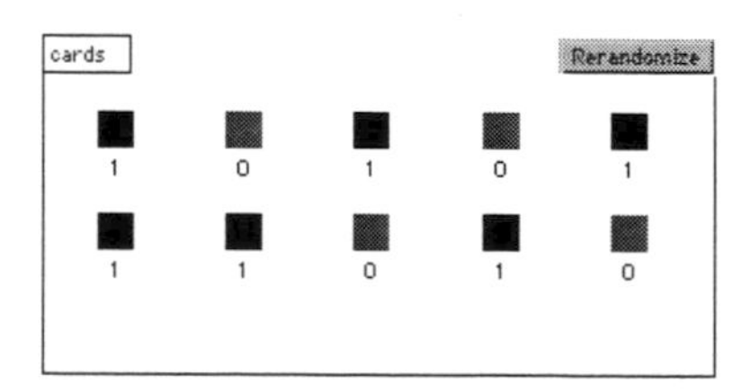

No gold balls! On the screen, you can see that the "zero" squares are red and the "1"s are black. Furthermore, we see the numbers 0 and 1, not the old "a case" caption.

- Double-click in that collection to open the inspector.
- Click on the **Display** tab to open that panel. It will look something like the illustration. You'll see a number of special attributes. In this case, two of them have formulas: **image** and **caption**. You can see all of **caption**'s formula; to see more of **image**'s formula, drag the border between **image** and **width** down as shown.

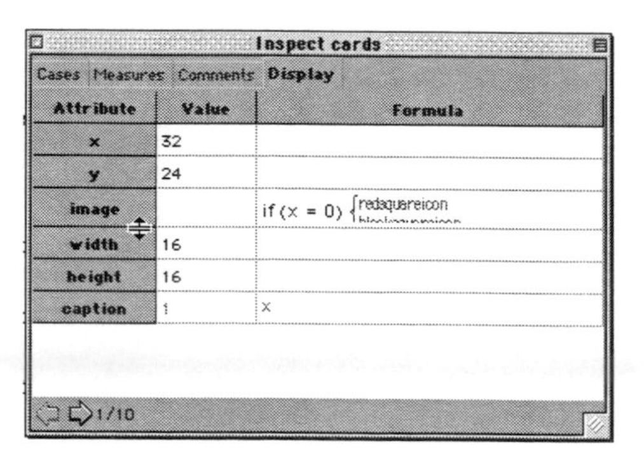

- To edit the formula for **image**, double-click it in the inspector; the formula editor opens. It should be clear how the **if( )** construction works. You can see the names of the icons, **redSquareIcon** and **blackSquareIcon**. Where did we get those?

Icon names are useful only in the **image** attribute in the **Display** panel.

- In the formula editor, on the right-hand side, is a list of categories of things—Attributes, Functions, Global Values—and Icon Names! Click the device to open up that list. You'll see something like the editor in the illustration.

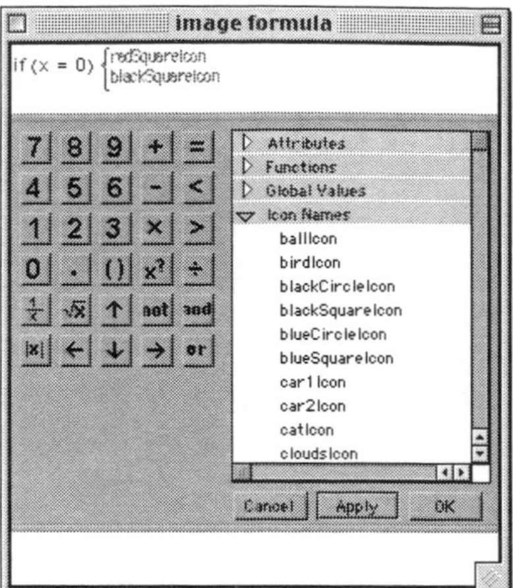

- Try editing the formula to get different icons. You can simply type the icon names you see into the formula, or double-click to insert them.
- Press **Apply** or **return** to see the results.
- Try changing the other attributes to see what happens. Some notes:
  - To change all of the cases, be sure to edit the formula for the display attribute. For example, to make the icons fatter, make a very simple formula (e.g., **32**) for **width**.
  - The attributes **x** and **y** are the positions of the *middle* of the icon in the collection. The **x** attribute is the horizontal coordinate, increasing to the right; **y** is vertical, increasing *as you go down.*
  - One of the most useful things you can do is give icons a more meaningful caption. Inspect the collection just below, **Measures from cards**, to see its **caption** formula, and the use of the **concat( )** function to build strings.

By controlling position and size, you can essentially build graphs of your own, though sometimes the formulas can get pretty arcane. Some of the demos (listed above), for example, build axes or backgrounds out of some cases, carefully ignoring them in calculations.

We have created another example of this sort of thing in **randomDisplayOnly.ftm**.

Controlling Sample Size with a Slider

See "Correlation Coefficients of Samples" on page 41 and "Standard Error and Standard Deviation" on page 66.

There are a number of sneaky ways to do this; in this book, we tend to use *filters*. A filter is a Boolean expression (that you enter using the formula editor, of course) that tells Fathom which cases to display in a graph or table. But if you apply the filter to the *collection*, it's as if the cases aren't there at all. They're hidden to all displays, and are not used in any calculations.

In addition, we use the special Fathom variable, **caseIndex**. That's the number of the current case. So we'll make a slider named **N**, and use **caseIndex < N** as our formula. Let's do it step by step:

- Open **ControllingSampleSize.ftm**. In it, you have a collection with one attribute, **a**, which is standard normal. There are 100 cases in the collection, displayed in a dot plot.

Now for the slider and filter.

- Drag a slider off the shelf. Name it **N**.

- Context-click the collection; choose **Add Filter**. The formula editor opens. Enter **caseIndex < N**. Close the formula editor with **OK**.

Suddenly, the graph loses almost all its points! Only 4 are left! That's because **N** is 5. In addition, the collection expands to show gold balls; you can see the filter at the bottom.

- Make the collection smaller by grabbing a corner and shrinking it until it becomes an icon.
- Play with the slider to change the number of points.

hold down **option** to get ≤

At this point, we would do different things depending on where we are headed. Just to make things less confusing, we might also, for example, use ≤ instead of <, or use **round(N)** instead of **N** in the filter.

A Different Way to Control Sample Size

See "The Central Limit Theorem" on page 96.

We used a different, more direct way to control sample size when we constructed the Central Limit Theorem demo. In that file (**Central Limit Theorem.ftm**), look at the **Sample** panel of the **sample** collection. It will look like the illustration. You can see that, instead of collecting a fixed number of cases, we sample until the **count( )** in this sample collection is greater than or equal to **N**, the slider.

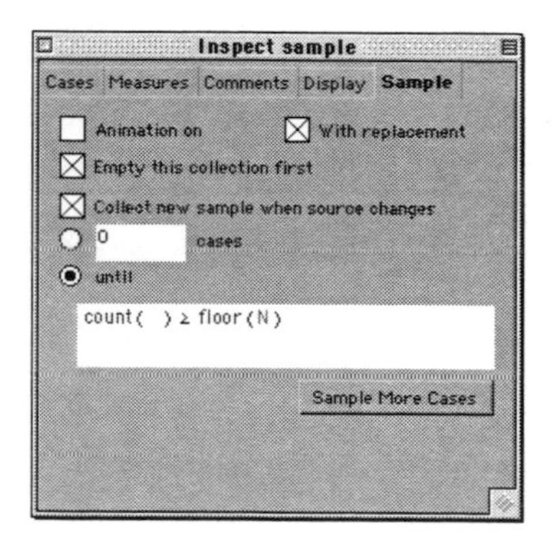

This sweet strategy does not work in the other files because there, we created "samples" not by sampling, but by using random number generators to *simulate* sampling from an infinite population. There was no **Sample** panel.

Simulating Correlated Data

See "Correlation Coefficients of Samples" on page 41 and "Regression Towards the Mean" on page 44

When we study correlation, it's useful to be able to generate data that has whatever correlation we want. A number of times in this book, we have a slider, **rho**, that represents the correlation in a population; then we display a sample from that population. As we drag **rho**, the data change to reflect the new correlation.

The math behind it goes like this: if x and h are standard normal, and if we construct y to be

$$y = \rho x + (\sqrt{1-\rho^2})h,$$

then y will be standard normal too, and the correlation of x and y will be ρ.[1]

Now you'll make a "correlation machine" of your own:

---

1. Why does this work? See "Correlated Data: Why the Way We Generate It Works" on page 177.

- Open the file **myCorrelationMachine.ftm**. The project is already started for you. You have a case table with 30 empty cases and three attributes: **x**, **h**, and **y**. You also have a slider named **rho**.
- Context-click the label—the column head—for **x**. Choose **Edit Formula**. The formula editor opens.
- Enter **randomNormal( )**. Close the editor with **OK**. Do the same for **h**. Random numbers appear in those two columns.
- Now give **y** a formula as shown in the illustration. Be sure to use an asterisk for multiplication between **x** and **rho** (or Fathom will look for an attribute named **xrho**.) Close the editor with **OK**.

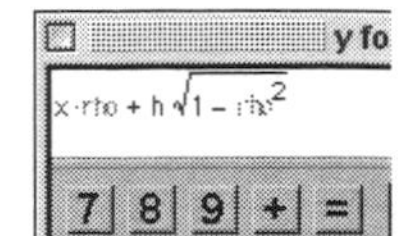

- Drag a graph off the shelf; put **x** on the horizontal axis and **y** on the vertical.
- You're done! Play with the **rho** slider!
- **Save** your work. You can use this, especially to impress people that a correlation as high as +0.5 doesn't look particularly correlated.

Enhancements

- Iconify the case table (drag its corner to make it small) since you don't need it any more.
- Make the graph larger. Make sure its bounds encompass at least –2.5 to +2.5 on each axis.
- Add a least-squares line.
- Rerandomize repeatedly and compare the **r^2** value on the graph with the value of **rho** (squared) to see how much the former varies.

Using a Point As a Controller

See "Why np>10 is a Good Rule of Thumb" on page 107.

Fathom's sliders give you control over constants—but only one constant at a time. In the case of "Why np>10 is a Good Rule of Thumb" on page 107, we want to control two constants—**n** and **p**—at once. (We could have done this demo with two sliders, but it would have been much duller.)

So we came up with a scheme for tricking Fathom into using a data point as a two-dimensional slider. In that file, we have **Collection 1** with one point. That point has two attributes, **sample_p** and **n**. The graph displays that point as a scatter plot.

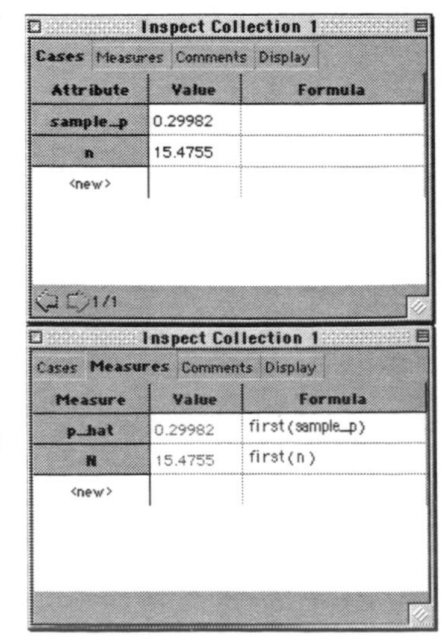

Now we have to use those positions in some other collection. So we use the "measures" mechanism. The first step is to make those values into measures so we can collect them. You can see how we do that in the illustration, which shows the **Cases** panel for **Collection 1** on the top, and its **Measures** panel on the bottom. These look almost identical—but the measures are determined by formula, using the function **first( )**, which returns the first—in this case, only—value in the collection.

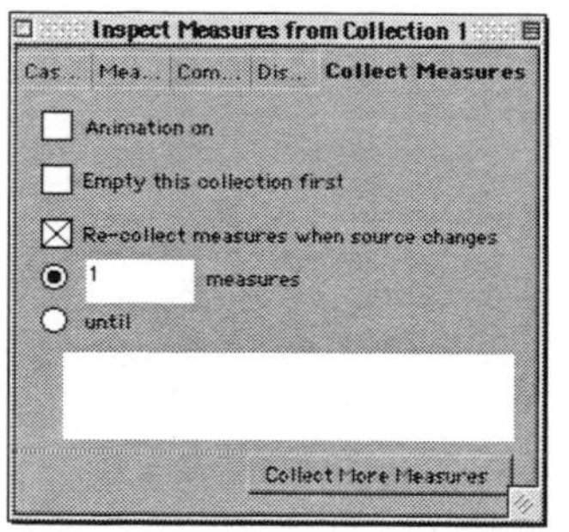

Then we collect measures, using an unusual setting (at left). We collect one point every time, we do not empty the collection first, and we collect whenever the source changes. This last setting is key: this way, every time we drag the point in the original collection, Fathom collects one more measure. If we keep dragging, it collects as fast as it can.

Then, in the measures collection, we do the calculation with **N** and **p_hat** to figure out whatever we need to about the CI; in this case, we create a new attribute, **goodCI**, that tells us whether the CI extends below zero. We use that attribute to code the points in the graph with different symbols. To make that work, drag the attribute name into the *middle* of the scatter plot. This makes **goodCI** what Fathom calls a "legend attribute." In a scatter plot, if the legend attribute is categorical, Fathom plots the points using different symbols; if it is continuous, Fathom plots the points as squares in different colors.

Using a Continuous Attribute as Categorical

See "Sampling Distributions and Sample Size" on page 85 and
"How the Width of the Sampling Distribution Depends on N" on page 88

By default, when you drag the name of a numeric attribute to the axis of a graph, you get a continuous axis. That means that every point shows up in the appropriate place on the axis, depending on its value. This seems obvious and normal, but that's not always what you want.

Sometimes you have numbers for data values, but you want to treat them as *categories*. An example is when we study how a distribution (of sample mean, say) depends on sample size. Then you may want parallel dot plots—or box plots or histograms—rather than a single scatter plot. To get that effect, make sure the **shift** key is down as you drop the attribute name.

The following figure shows you the difference. Notice how in the right-hand figure—probably not what you want—the **N** axis is a true numeric axis, where distances correspond to values.

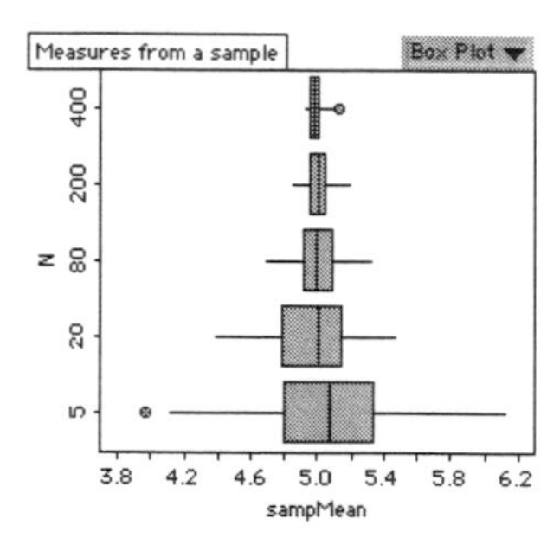

Here, when we dropped **N**, the **shift** key was down. We have parallel box plots.

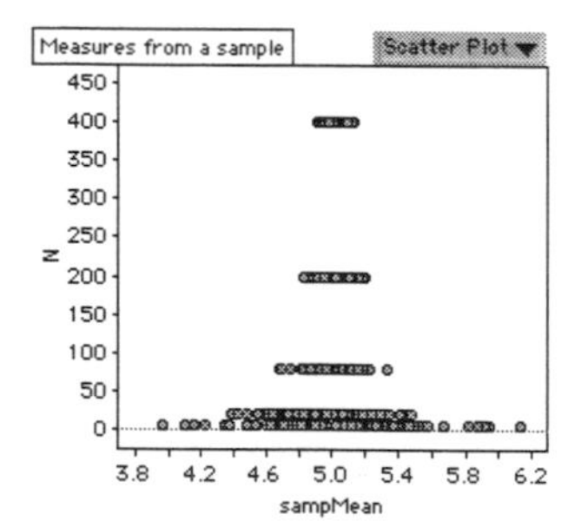

Here, the **shift** key was not down. We have a scatter plot.

In day-to-day data analysis, you also need this effect when you enter truly discrete data as numeric, e.g., if you have "room number" or (importantly) Likert survey responses.

Plotting Binomial Probability—and Other Discrete Distributions

See "The Confidence Interval of a Proportion" on page 100.

It's great that Fathom lets us plot curves on top of data, and most of the time it's really easy. But plotting discrete probability distributions—such as the binomial—is a little tricky; if you did "The Confidence Interval of a Proportion" on page 100, you had to type in a truly arduous function to get a theoretical probability distribution to show up.

This section explains why, and gives you some basic help.

First of all, in the formula editor, you can learn about a function by selecting it in the hierarchical list on the right side of the window. You may have to open several lists to find the one you want. But when you select it (clicking once) on the right, some help appears in the bottom pane. (All of these panes are resizable.) The illustration shows how we opened **Functions**, then **Distributions** within that, then **Binomial**, to find the help for **binomialProbability**.

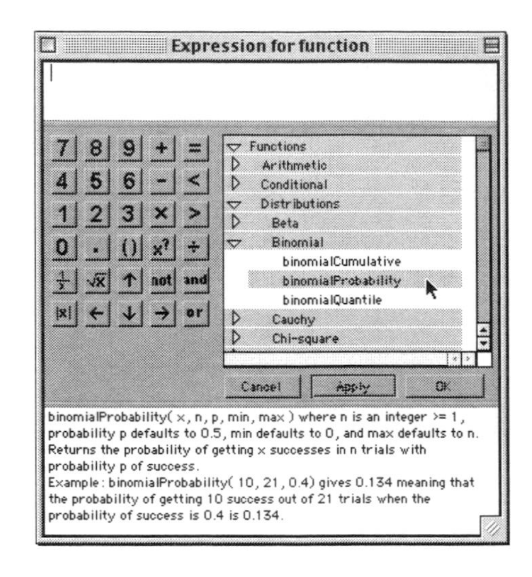

Ignore **min** and **max**, and focus on the example: **binomialProbability(10, 21, 0.4)** is the probability of getting (exactly) 10 successes in 21 trials when the probability of success is 0.4.

The key thing for plotting is to note that unless the first argument is a whole number, *the probability will be zero*. For example, there is no chance of getting exactly 9.8 successes. So when you plot it, Fathom will give you a line at zero—not what you want. What we want is usually a stairstep function that looks kind of like a histogram, where the probability of getting 4 successes is displayed for the whole range from 3.5 to 4.5.

So you use Fathom's **round( )** function, like this:

binomialProbability(round(x), round(N), p)

where, in this case, **N** and **p** are sliders. Note, that since the **N** parameter has to be an integer, you need to use **round( )** (or **floor( )** or **trunc( )**) on **N** as well.

Look at the file **How to Plot Binomial Prob.ftm** (in the **other files** folder) for an example.

Why do we have to jump through hoops to get this "stairstep" display of a probability function? Until Fathom can better read our minds, we should remember that it is perpetrating a lie: the function really has zero values except at the integers. We are finessing values that are true probabilities to look like probability *densities*—probabilities *per unit on the other axis*. This also explains the function names: discrete distributions, whose domains are integers, are named with "probability," e.g., **binomialProbability** and **geometricProbability**. The continuous distributions are named with "density," e.g., **normalDensity** and **tDensity**.

A Little Mathematical Statistics

Fifty Fathoms lets us witness the details of statistics up close, using empirical techniques: dragging values, generating random numbers, resampling, doing repeated tests, etc. But there is another approach, more abstract, more mathematical, and quite elegant. It is through mathematical statistics that we learn, for example, why the denominator in the sample variance is not $(n - \pi/3)$, which works pretty well, but instead $n - 1$, which is exactly correct.

Throughout this appendix, I have been aided immeasurably by Michael Allwood of Brunswick School in Greenwich, Connecticut, whose expertise in this far exceeds my own. Michael not only reviewed, edited, and reshaped the text and formal derivations to make them clearer and more correct; he also suggested completely different (i.e., better) approaches. Beyond that, he has taught me much that I had learned and forgotten, and much that I had never seen. A large portion of the intellectual effort in the next twelve pages is his; the remaining sloppiness is mine.

The derivations that follow help us get at one of the central ideas of inferential statistics: learning about a *population* from the data in a *sample*. Many of the demos in this book are about populations and samples; typically, in these demos, we know the population parameters and get to see the distribution of sample statistics as we repeatedly sample, often with various sample sizes. When we actually use statistics, though, we get only one sample, and never know the population parameters. How do we connect these two situations? To answer this, we need to discuss *random variables*.

Some Basics

To discuss random variables, we need to talk about *expectation* and *variance*. Almost everything that follows derives from properties of these two concepts.

Let's use an example to get some definitions and notation straight.

Suppose we plan to measure a (randomly-chosen) student's armspan. Let's call that number X. This thing is a *random variable*, because we don't know what value it will have. We traditionally use capital letters such as X, Y, or H to name random variables.

Note how this X is different from other variables, such as the x in $y = mx + b$. We don't know the value of this x, either. In fact, it doesn't make sense to ask for the "value" of x. It can take on *all* real values; we use it to help describe a function. Alternatively, consider the meaning of x in $3x^2 - \sin(2x) = 0$. Here, x is fixed (perhaps with many values) but unknown—because we haven't solved the equation. But at least we would get the same answer every time we solved it. Our new X is more substantially unknown, because if we *repeat* the effort to determine X, it might take on a different value.

Let us actually measure a student's armspan. We get a sample value, x_1, the armspan of that student (1.65 meters). Here we use lower case to indicate that this is a sample value—something that has been actually done. Big X, the random variable, is the subjunctive "if I were to measure an armspan…" rather than the indicative "I measured it."

If we keep measuring armspans and keep track of them, we can get many little *x*'s: $\{x_1, x_2, x_3, \ldots, x_i, \ldots, x_n\}$. This is a set of sample values of *X*. We can compute sample statistics for these *n* values, for example, using this formula for the sample mean:

$$\bar{x} = \frac{1}{n}\sum_{i=1}^{n} x_i,$$

This is almost the root-mean-square deviation from the mean except for the $n-1$ in the denominator instead of *n*.

and this one for the sample standard deviation *s*:

$$s = \sqrt{\frac{1}{n-1}\sum_{i=1}^{n}(x_i - \bar{x})^2}.$$

(Why lower case? Because these represent actual values. We can't compute the sample statistics until we have the armspans. Why $n-1$ rather than *n* in the denominator? More about this later.)

These sample statistics $\bar{x}$ and *s* tell us about the sample. But they also give us an idea about the population—about the armspans of all possible students. We will find out that $\bar{x}$ is an unbiased estimator for the population mean μ, and that s^2 is an unbiased estimator for the population variance, σ^2. So let's take a few pages to consider the population.

If *X* is a student's armspan, we define the *expectation* or *expected value* of *X*, *E*(*X*), to be the *mean* armspan of all possible students. This is the *population mean*. That is, $E(X) = \mu$.

We'll define the variance, Var(*X*), as the expectation of the squared deviation of *X* from the population mean. That is, $\text{Var}(X) = E[(X-\mu)^2]$. Now that we have the variance, we can define the population standard deviation $\text{SD}(X) = \sigma = \sqrt{\text{Var}(X)}$.

What happens to expectation and variance when we transform or combine random variables? We'll start out from definitions, and then use early results to find later ones.

Expectation

We said that $E(X) = \mu$. More precisely, we define the expected value of a random variable *X* to be

$$E(X) = \sum x\text{P}(X = x), \qquad \text{Equation 1}$$

where the summation is over all possible values *x* of the random variable *X*, and where $\text{P}(X = x)$ is the probability that a particular value *x* of the random variable occurs. We'll also write $\text{P}(X = x)$ as $\text{P}(x)$.[1]

This is not as scary as it looks. For example, the expectation of the sum of two fair dice is the sum over all possible values {2, 3, 4, …, 12} times the probability of each:

$$E(X) = 2\times\frac{1}{36} + 3\times\frac{2}{36} + \ldots + 7\times\frac{6}{36} + \ldots + 12\times\frac{1}{36}$$
$$= 7.$$

In a simpler case, if the population comprises *n* equally-likely values, each P(*x*) is simply $1/n$.

1. I like to think of Equation 1 as telling us that the expected value is a weighted mean, *weighted by probability*.

This generalizes readily to the continuous case (which we will not treat here in detail), where you use integrals instead of summations. The formula for $E[X]$—the population mean—uses a probability density function $f(x)$ in place of the probability $P(x)$, like this:

$$\mu = \int_{-\infty}^{\infty} xf(x)dx \text{, where } f(x) \geq 0 \text{ for all } x \text{ and } \int_{-\infty}^{\infty} f(x)dx = 1 .$$

Let's figure out what happens to the expectation when we change a random variable using a linear transformation:

$$\begin{aligned} E[aX+b] &= \sum (ax+b)P(x) \\ &= \sum axP(x) + \sum bP(x) \\ &= a\sum xP(x) + b\sum P(x) \\ &= aE[X] + b. \end{aligned}$$

Equation 2

This "transformation formula" mirrors the familiar result that if you add a constant to a data set, its mean increases by that amount; and if you multiply each value by a constant, the mean multiplies as well. Notice how we used $\sum P(x) = 1$, because the sum is over *all* possible values x of X.

For a continuous random variable, the result in Equation 2 is the same, and the proof is very similar to the above, although we integrate rather than sum.

Expectations have other important properties, such as

$$E(X+Y) = E(X) + E(Y),$$

Equation 3

which may seem obvious, but is a little tricky to prove, and

$$E(XY) = E(X)E(Y),$$

Equation 4

which is true if X and Y are independent.

Variance

Let's develop the transformation formula for variance.

$$\begin{aligned} \text{By definition, } \mathrm{Var}(X) &= E[(X-\mu)^2]. \\ \text{So } \mathrm{Var}(aX+b) &= E[(aX+b-a\mu-b)^2] \\ &= E[(a(X-\mu))^2] \\ &= a^2E[(X-\mu)^2] \text{, that is,} \end{aligned}$$

$$\mathrm{Var}(aX+b) = a^2\mathrm{Var}(X).$$

Equation 5

Remember that variance is the square of the SD; this mirrors the familiar result with data that adding a constant to each value does not change the SD, but multiplying increases the SD by that factor.

Now we'll derive an alternative formula for variance:

$$\begin{aligned}\mathrm{Var}(X) &= E[(X-\mu)^2] \\ &= E[(X^2 - 2X\mu + \mu^2)] \\ &= E(X^2) - 2\mu E(X) + \mu^2 \\ &= E(X^2) - 2\mu^2 + \mu^2 \\ &= E(X^2) - \mu^2.\end{aligned}$$

That is,

$$\mathrm{Var}(X) = E(X^2) - [E(X)]^2. \qquad \text{Equation 6}$$

Thus the variance is the mean of the squares minus the square of the mean.

Now we will use this alternative formula for variance, Equation 6, to derive the all-important addition formula for variance. If X and Y are independent (so we can use Equation 4):

$$\begin{aligned}\mathrm{Var}(X+Y) &= E[(X+Y)^2] - [E(X+Y)]^2 \\ &= E[X^2 + 2XY + Y^2] - [\mu_X + \mu_Y]^2 \\ &= E[X^2] + 2\cdot E[X]E[Y] + E[Y^2] - \mu_X^2 - 2\mu_X\mu_Y - \mu_Y^2 \\ &= \{E[X^2] - \mu_X^2\} - \{E[Y^2] - \mu_Y^2\} + 2\mu_X\mu_Y - 2\mu_X\mu_Y.\end{aligned}$$

But $\{E[X^2] - \mu_X^2\} = \mathrm{Var}(X)$ and $\{E[Y^2] - \mu_Y^2\} = \mathrm{Var}(Y)$.

So if X and Y are independent,

$$\mathrm{Var}(X+Y) = \mathrm{Var}(X) + \mathrm{Var}(Y). \qquad \text{Equation 7}$$

The Distribution of the Sample Mean

We're often concerned with samples, and we frequently collect sample means and look at their distribution. So let's look at some properties of the distribution of the sample mean from this math-stats point of view.

The mean of a sample of size n drawn from random variable X is $\bar{x}$. This is probably not μ, and it will be different for every sample. That is, $\bar{x}$ is *not a constant* in this context, but rather a random variable different from X—though their distributions are related.

Some may still call this sample mean $\bar{x}$, but we'll reserve that for a situation where we already have the sample values.

From here on, since it's a random variable, we'll call the sample mean $\bar{X}$, with a capital X. Similarly, when we imagine future samples, we don't know what the first element[1] in the sample will be—so this first element is also a random variable. So the individual items in a sample-to-be are the random variables $X_1, X_2, X_3, \ldots X_n$. Each one of these is an instance of X itself, so it has the same distribution. In particular, since $E(X) = \mu$, each $E(X_i) = \mu$ as well.

The expected value of $\bar{X}$ is

1. in Fathom, the first *case*

$$E(\bar{X}) = E\left(\frac{X_1 + X_2 + \ldots + X_n}{n}\right)$$
$$= \frac{1}{n}(E(X_1) + E(X_2) + \ldots + E(X_n))$$
$$= \frac{1}{n}(\mu + \mu + \ldots + \mu)$$
$$= \mu.$$

This is what we mean when we say that $\bar{X}$ is an *unbiased estimator* of μ.

Digression

It is not at all surprising that we expect the sample mean to hover around the population mean, so why include this derivation? On one level, to show how Real Mathematical Statisticians treat these issues: how $\bar{X}$ is a random variable; how we can use what we've learned to treat each X_i as a random variable; how careful we are not to overstep what we have derived. On another level, this reminds us how treacherous these things are. We are about to derive non-obvious things. How will we know we got them right? One answer is at the core of this book: we can check them out through simulation. In this case, we can take a thousand samples and see if the distribution of sample means centers around the population mean.

It does, and a good thing, too.

Onward!

That's the center; the spread is harder. We will use some of the variance properties we derived above; we will work with variance, rather than SD, to avoid hairy roots. Here we go:

$$\mathrm{Var}(\bar{X}) = \mathrm{Var}\left(\frac{X_1 + X_2 + \ldots + X_n}{n}\right)$$
$$= \frac{1}{n^2}[\mathrm{Var}(X_1 + X_2 + \ldots + X_n)].$$

But the X_i's are independent, so by Equation 7 on page 174,

$$\mathrm{Var}(\bar{X}) = \frac{1}{n^2}[\mathrm{Var}(X_1) + \mathrm{Var}(X_2) + \ldots + \mathrm{Var}(X_n)]$$
$$= \frac{1}{n^2}[n\mathrm{Var}(X)].$$

That is,

$$\mathrm{Var}(\bar{X}) = \frac{\sigma^2}{n}. \qquad \text{Equation 8}$$

Thus, the variance of the sampling distribution of the mean is $1/n$ times the variance of the source distribution. We have made no assumptions about the distribution itself, except that the variance exists. It need not be normal. Note: this is the *variance of the sample mean*—the spread of the distribution of sample means—not the *sample variance*—an estimator of the variance of the population. We'll get to that soon in "Sample Variance: Why the Denominator is n – 1" on page 178.

A Random Walk: Two Proofs That the Mean Square Distance is N

The previous section was background. Now let's look at specific topics that have come up in the book. We begin with what we asserted in "How Random Walks Go as Root N" on page 51: that in a simple, one-dimensional random walk—using steps +1 and −1, with equal probability—the mean square distance from the endpoint to the origin is equal to the number of steps.

This proof is entirely due to Michael Allwood.

Let's let each step be a random variable X which takes on the values +1 and −1, each with probability 1/2. By Equation 1 on page 172, $E[X] = 0$; using Equation 6 on page 174, we find that $\text{Var}(X) = 1$.

Now the absolute distance from the origin after n steps is

$$D = \left|\sum_{i=1}^{n} X_i\right|, \text{ and the squared distance is } D^2 = \left[\sum_{i=1}^{n} X_i\right]^2.$$

Let's define Y to be the "signed" distance from the origin,

$$Y = \sum_{i=1}^{n} X_i.$$

Then, from the addition of expectations, $E[Y] = 0$; and from addition of variances, $\text{Var}(Y) = n$.

We're interested in the *mean square* distance, $E[D^2]$—which is also $E[Y^2]$. To get that (we have no rule for the expectation of a product when the factors are not independent) we can use the variance formula Equation 6:

$$\begin{aligned} \text{Var}(Y) &= E[Y^2] - [E(Y)]^2 \\ n &= E[Y^2] - 0 = E[D^2]. \end{aligned}$$

That is,

$$E[D^2] = n,$$

which is what we wanted to show: the mean square distance from the origin after n steps of a random walk is just n, the number of steps.

An Alternate Proof

Suppose you don't believe this random-variable stuff and want a good old-fashioned inductive proof. Here is one. As usual, we show that the assertion (the mean square distance from the endpoints of all possible walks to the origin is n) is true for $n = 1$, and then take the inductive step, assuming it true for some value k and showing that it must then be true for $n = k + 1$:

Do One First. With a one-step walk, the only possible distance we can go is one step. So the mean distance is one, and so is the mean square distance. And that's n, the number of steps.

Inductive Step. Suppose that there are m possible k-step walks, and their endpoints are $\{e_1, e_2, e_3, \ldots, e_m\}$. For our inductive step, we assume that the mean square distance from these endpoints to the origin is k. That is, our assumption is that the mean square distance after k steps is

$$MSD_k = \frac{1}{m}\sum_{i=1}^{m} e_i^2 = k. \qquad \text{Equation 9}$$

(In fact, $m = 2^k$, and the probability of each endpoint is $1/2^k$, but we don't need that for this proof.)

Now in a $(k+1)$-step walk, we notice that for every k-step ending position e_i, there are two possibilities for the next endpoint: $e_i + 1$ and $e_i - 1$. Also, the number of $k+1$-step walks is $2m$. So the mean square distance for a $(k+1)$-step walk is

$$\begin{aligned} MSD_{k+1} &= \frac{1}{2m}\left[\sum_{i=1}^{m}(e_i+1)^2 + \sum_{i=1}^{m}(e_i-1)^2\right] \\ &= \frac{1}{2m}\left[\sum_{i=1}^{m}(e_i^2+2e_i+1) + \sum_{i=1}^{m}(e_i^2-2e_i+1)\right] \\ &= \frac{1}{2m}\left[\sum_{i=1}^{m}(e_i^2+e_i^2) + \sum_{i=1}^{m}(2e_i-2e_i) + \sum_{i=1}^{m}(1+1)\right] \\ &= \frac{1}{m}\left[\sum_{i=1}^{m}e_i^2 + \sum_{i=1}^{m}1\right] \\ &= \frac{1}{m}\sum_{i=1}^{m}e_i^2 + \frac{1}{m}(m) \\ &= k+1. \end{aligned}$$

This completes the inductive proof. (The last step uses Equation 9.)

Correlated Data: Why the Way We Generate It Works

Thanks to Matt Litwin, Chris Olsen, and Dick Scheaffer for help on this section.

In "Correlation Coefficients of Samples" on page 41 and "Regression Towards the Mean" on page 44, we use a slider **rho** to control the underlying correlation between two attributes. In "Simulating Correlated Data" on page 167, we describe how we set that up. This section describes why that formula works.

If X and Y are random variables, we define their correlation coefficient ρ (Greek letter rho) to be

$$\rho = \frac{E[(X-\mu_X)(Y-\mu_Y)]}{\sqrt{\mathrm{Var}(X)\mathrm{Var}(Y)}}.$$

But if X and Y are standard normal, $\mu_X = \mu_Y = 0$ and $\mathrm{Var}(X) = \mathrm{Var}(Y) = 1$. So

$$\rho = E[XY],$$

that is, the correlation coefficient of any two standard normal random variables X and Y is the expected value of their product. Bear that in mind; we'll be back to it shortly.

Now think about two independent standard normal variables, X and H (for helper). We're going to make a linear combination of them to get Y, like this:

$$Y = aX + bH. \qquad \text{Equation 10}$$

Y is normal because a linear combination of independent normal random variables is normal. (We will not prove this.) But if we want Y to be *standard* normal as well, what do we know about a and b? First, since the means of X and H are both zero, the mean of Y is also zero. (This does not help with a and b, but it's necessary if Y is to be standard normal.)

Next, for Y to be standard normal, we must make sure that $\mathrm{Var}(Y) = E[Y^2] = 1$. So we square the above, take its expected value, and do some algebra:

$$\begin{aligned} E[Y^2] &= E[(aX+bH)^2] \\ 1 &= E[a^2X^2 + 2abXH + b^2H^2] \\ 1 &= a^2E[X^2] + 2abE[XH] + b^2E[H^2]. \end{aligned}$$

Now, X and H are independent and their means are zero, so $E[XH] = 0$. And since X is standard normal, $\mathrm{Var}(X) = E[X^2] = 1$; similarly for H. So

$1 = a^2 + b^2$, which means that

$$b = \sqrt{1 - a^2}.$$

So Y will be standard normal as long as a and b are related in that Pythagorean way.

Now let's relate a to ρ, the correlation coefficient of X and Y—which is the expected value of their product:

$$\begin{aligned} E[XY] &= E[X(aX+bH)] \\ \rho &= E[aX^2] + E[bXH] \\ &= a \times 1 + b \times 0 \\ &= a, \end{aligned}$$

which gives us what we were looking for. Equation 10 becomes $Y = \rho X + \sqrt{1-\rho^2}H$.

Sample Variance: Why the Denominator is n – 1

In "Does n – 1 Really Work in the SD?" on page 91, we saw how the formula for standard deviation with $n - 1$ in the denominator is not an unbiased estimator of the true standard deviation—but that the formula for sample *variance* with $n - 1$ in the denominator *is* an unbiased estimator of the population variance.

But why $n - 1$? We will first discuss conceptually why it's less than n, and then we'll show it's exactly $n - 1$ using some algebra. In this effort, we are greatly aided by two contributors to the AP Statistics list-serve—Steve Gold and Michael Allwood—whose crystalline posts in the Fall of 2001 helped me get clear on this.

Suppose we have a population (which need not be normally distributed) with mean μ and standard deviation σ. We draw a sample from that population, and that sample has a mean of $\bar{x}$. Suppose we calculate the mean square distance from the true population mean, like this:

$$V = \frac{1}{n}\sum(x_i - \mu)^2.$$

Now it turns out that $E[V] = \sigma^2$. It would be great to use V as an unbiased estimator of σ^2, but we have a problem: we cannot calculate it, because we do not know the population mean μ. The best we can do is to use $\bar{x}$. If we calculate this mean square distance using $\bar{x}$ (let's call it Λ), we have

$$\Lambda = \frac{1}{n}\sum(x_i - \bar{x})^2.$$

How can we relate Λ to V (and ultimately to σ^2)? If you have done "The Mean is Least Squares, Too" on page 30, you know that $\bar{x}$ is the value that makes Λ a minimum. That is, any other value (e.g., μ) will make that sum larger. Put another way, using $\bar{x}$ instead of μ makes us systematically *underestimate* the true variance. $\Lambda \le V$, and $E[V] = \sigma^2$. This makes sense: the points in our sample will tend to be closer to their own mean than to that of the population.

Our unbiased estimator, then, has to be *bigger* than Λ. How will we find it?

Saying It with Symbols

We start with the formula for the variance that appears as Equation 6 on page 174:

$$\mathrm{Var}(X) = E(X^2) - E(X)^2.$$

We can rearrange it (and change some notation) to get

$$E(X^2) = \sigma^2 + \mu^2. \quad \text{Equation 11}$$

This is true for *any* random variable. Therefore, since $\bar{X}$ (the sample mean) is also a random variable, and it has mean μ and variance σ^2/n (as we found in Equation 8 on page 175), we can write:

$$E(\bar{X}^2) = \frac{\sigma^2}{n} + \mu^2. \quad \text{Equation 12}$$

Hold that thought. When we calculate a sample variance, we will not have μ but rather $\bar{X}$. With our data, we can actually calculate

$$\sum(X_i - \bar{X})^2, \text{ not } \sum(X_i - \mu)^2.$$

So let us study that first expression. First, we do some algebra:

$$\begin{aligned}\sum(X_i - \bar{X})^2 &= \sum(X_i^2 - 2X_i\bar{X} + \bar{X}^2) \\ &= \sum X_i^2 - 2\bar{X}\sum X_i + n\bar{X}^2 \\ &= \sum X_i^2 - 2\bar{X}(n\bar{X}) + n\bar{X}^2 \\ &= \sum X_i^2 - n\bar{X}^2.\end{aligned}$$

We want to know how this quantity would behave if we were to sample repeatedly. So we cast this as a statement about expected values:

$$\begin{aligned} E[\sum (X_i - \bar{X})^2] &= E(\sum X_i^2) - nE(\bar{X}^2) \\ &= \sum E(X_i^2) - nE(\bar{X}^2) \\ &= n(E(X^2) - E(\bar{X}^2)). \end{aligned}$$

In the last step, we can lose the subscripts because X_i has the same distribution as X. Now we substitute in the results from Equation 11 and Equation 12:

$$\begin{aligned} E[\sum (X_i - \bar{X})^2] &= n(E(X^2) - E(\bar{X}^2)) \\ &= n\left(\sigma^2 + \mu^2 - \left(\frac{\sigma^2}{n} + \mu^2\right)\right) \\ &= (n-1)\sigma^2. \end{aligned}$$

So

$$E\left[\frac{\sum (X_i - \bar{X})^2}{n-1}\right] = \sigma^2,$$

which is just what we mean when we say that the expression in brackets is an *unbiased estimator* of the population variance.

The Geometric Distribution: Proof That the Mean is (1/p)

The geometric distribution gives you whole numbers—the number of trials it takes until some event happens. The probability that it happens in one trial is p. The probability that it happens eventually is 1. But what is the *mean* or *expected* number of trials? We looked at this experimentally in "Wait Time and the Geometric Distribution" on page 154. What's the theoretical result?

For example, suppose our event is rolling a three with a fair die. Here's one way to think of it: suppose you roll the die 6000 times. We expect 1000 threes in those 6000 rolls; so on average, it takes 6 rolls for each three to come up.

How to prove it? Let the random variable K be the number of rolls it takes to get a three. We have a 1/6 chance of getting it in one trial. If we don't get it—chance 5/6—we expect it to take $E[K]$ more rolls, for a total expected number of $E[K] + 1$. So

$$E[K] = \frac{1}{6} + \frac{5}{6}(E[K] + 1), \text{ which leads to } E[K] = 6.$$

Generalizing, $E[K] = p + (1-p)(E[K]+1)$, or $E[K] = 1/p$

If that is not convincing, try this: To calculate its expected value, we use Equation 1 on page 172. We will calculate the probability of each result (each number of steps), multiply the probability by that number of steps, and add up all those products. Since the process could go on forever, the sum will have an infinite number of terms.

The chance that we get a three on the first roll (i.e., $P(K = 1)$)is $1/6$. The chance that it happens for the first time on the second roll is $(5/6)(1/6)$. The 5/6 is the chance that we *didn't* get it on the first roll. So the contribution that "one" makes to the average is (1)(1/6); two's contribution is (2)(5/6)(1/6); and so on.

The probability that you roll a three for the first time on the *k*th roll is

$$P(K = k) = \left(\frac{1}{6}\right)\left(\frac{5}{6}\right)^{k-1}. \text{ For example, } P(K = 5) = \left(\frac{1}{6}\right)\left(\frac{5}{6}\right)^{4}.$$

So the sum we have to add is

$$E[K] = \sum_{k=1}^{\infty} P(K = k) \times k = \frac{1}{6} \times 1 + \frac{1}{6}\left(\frac{5}{6}\right) \times 2 + \frac{1}{6}\left(\frac{5}{6}\right)^2 \times 3 + \ldots.$$

To find the sum of this infinite series, we're going to employ a trick: we'll multiply the whole thing by (5/6) and subtract. This will get us most of the way there.

$$E[K] = \frac{1}{6} \times 1 + \frac{1}{6}\left(\frac{5}{6}\right) \times 2 + \frac{1}{6}\left(\frac{5}{6}\right)^2 \times 3 + \ldots$$

$$\left(\frac{5}{6}\right)E[K] = \frac{1}{6}\left(\frac{5}{6}\right) \times 1 + \frac{1}{6}\left(\frac{5}{6}\right)^2 \times 2 + \ldots$$

Subtracting gives

$$\left(\frac{1}{6}\right)E[K] = \frac{1}{6} \times 1 + \frac{1}{6}\left(\frac{5}{6}\right) \times 1 + \frac{1}{6}\left(\frac{5}{6}\right)^2 \times 1 + \ldots$$

$$E[K] = 1 + \left(\frac{5}{6}\right) + \left(\frac{5}{6}\right)^2 + \left(\frac{5}{6}\right)^3 + \ldots.$$

Now $E[K]$ is a regular old geometric series. Assuming we don't remember the formula for the sum of an infinite geometric series, we use the same trick:

$$E[K] = 1 + \left(\frac{5}{6}\right) + \left(\frac{5}{6}\right)^2 + \left(\frac{5}{6}\right)^3 + \ldots$$

$$\left(\frac{5}{6}\right)E[K] = \left(\frac{5}{6}\right) + \left(\frac{5}{6}\right)^2 + \left(\frac{5}{6}\right)^3 + \left(\frac{5}{6}\right)^4 + \ldots.$$

This time, subtracting gives

$$\left(\frac{1}{6}\right)E[K] = 1,$$

$$\text{so } E[K] = 6.$$

That is, the mean number of rolls you need to roll a three is 6.

Now, to generalize. Suppose the probability of success on any given trial is p. Then $q = 1 - p$ is the chance that it does *not* happen on that trial.

The probability that it will happen for the first time on the *k*th trial is

$$P(K = k) = pq^{k-1},$$

so the sum we have to add—using Equation 1 to get expected value—is

$$E[K] = S = \sum_{k=1}^{\infty} kpq^{k-1}.$$

To add this infinite series S, we're going to employ the same tricks:

$$S = p + 2pq + 3pq^2 + \ldots$$

$$qS = \quad pq + 2pq^2 + 3pq^3 + \ldots$$

Subtracting gives

$$(1-q)S = p + pq + pq^2 + \ldots$$

$$pS = p + pq + pq^2 + \ldots$$

$$S = 1 + q + q^2 + q^3 + \ldots$$

Now we multiply through by q:

$$qS = \quad q + q^2 + q^3 + q^4 + \ldots$$

Subtracting again gives

$$pS = 1,$$

$$\text{So } E[K] = \frac{1}{p}.$$

Normally Distributed Random Numbers: How Do We Generate Them?

This is a small thing that we will not prove, but it's so useful it's worth knowing about.

Most random number generators simulate pulling numbers from a uniform distribution. Frequently we would like numbers pulled from a *normal* distribution. We can change the former into the latter using what's called a Box-Muller transformation[1].

Begin with two random numbers X_1 and X_2, independently drawn from a uniform distribution in the range [0, 1]. (This is the same as Fathom's **random( )** function.) Then create Z_1 and Z_2 like this:

$$Z_1 = \sqrt{-2\ln X_1}\cos(2\pi X_2)$$

$$Z_2 = \sqrt{-2\ln X_1}\sin(2\pi X_2)$$

Then Z_1 and Z_2 are both normally distributed with mean $\mu = 0$ and variance $\sigma^2 = 1$. The proof of this is too complex to look at here. But to get an idea of how the transformation works, open **BoxMuller.ftm** and select swaths of points in one graph or the other to see how they map. You will see, for example, that small values of **X1** give points that are far from the origin in the **Z1-Z2** plane.

---

1. G.E.P. Box and M.E. Muller. "A note on the generation of random normal deviates." *Ann. Math. Statistics*, v29, 1958, 610–611. George E. P. Box famously said: "All models are wrong, but some are useful." See also **http://mathworld.wolfram.com/Box-MullerTransformation.html**.

Acknowledgements

I would especially like to thank Michael Allwood, Jill Binker, Kristen Clegg, Bill Finzer, Meg Holmberg, Paul L. Myers, Chris Olsen, Tony Thrall, and three anonymous reviewers for their careful readings of the manuscript and for their support during the process of writing this book. They found lots of mistakes; the ones that remain are all mine—but I will publish errata on the web site (http://www.eeps.com).

I am also grateful to the Fathom programming team—Bill Finzer, Matt Litwin, and Kirk Swenson—for making the version of Fathom that comes with this first printing, helping figure out the CD, giving the documents their special identity, and being, as usual, cheery, interested, and frighteningly competent.

I also thank the staff of KCP Technologies, who generously lent me a machine so I could reformat every last document so it would look right under Windows; Richard Bonacci and his staff at Key College Publishing, who found reviewers and were extremely encouraging; and of course, Steve Rasmussen, who helps lots of us tyros figure out how or whether to publish anything. His enthusiasm is peerless and invaluable.

Tim Erickson
Oakland, California
September 2002

For Meg and Anne,
whose love and support—
unlike most quantities of interest to statisticians—
do not vary.